MW00714736

WHIZ-BANG DEMONSTRATIONS

This book was printed with soy-based ink on acid-free recycled content paper, containing 10% POSTCONSUMER WASTE.

HOLT, RINEHART AND WINSTON

A Harcourt Classroom Education Company

Austin • New York • Orlando • Atlanta • San Francisco • Boston • Dallas • Toronto • London

To the Teacher

There's no better way to help students learn a science lesson than to capture their attention from the very beginning. With 65 exciting demonstrations, this booklet offers you many opportunities to "wow" your students. Whether you are introducing a new chapter or reinforcing a difficult concept, these demos are sure to motivate.

Each of the demonstrations in this booklet was written with the teacher in mind. Although some demos involve student participation and others could be converted into a student activity, most are designed to be quick motivators performed by the teacher. Each demonstration contains useful information such as the following:

Purpose	What to Do
Time Required	Explanation
Advance Preparation	Discussion Questions
Helpful Hints	Going Further Extension Activity

The demonstrations in this booklet are organized into the following three sections—Life Science, Earth Science, and Physical Science. At the top of every instruction sheet you will see one of the following icons: Discovery Lab or Making Models. These icons indicate whether the purpose of the demo is to encourage students to discover the answer to a question on their own or to use a scientific model. At the end of many demonstrations, Critical Thinking and Going Further strategies are available for helping students process and apply the concepts presented.

CLASSROOM TESTED & APPROVED

You will also notice this icon, which acknowledges the many teachers around the country who helped ensure the safety, accuracy, and enjoyment of these demonstrations.

Art Credits
All art, unless otherwise noted, by Holt, Rinehart and Winston.
Abbreviated as follows: (t) top; (b) bottom; (l) left; (r) right; (c) center; (bkgd) background.
Front cover (owl), Kim Taylor/Bruce Coleman, Inc.; Page 2 (cr), Mazer Corp.; 3 (tl), Mazer Corp.; 6 (cr), David Kelley; 8 (bl), Kiwi Studios; 9 (tc), Layne Lundstrom; 12 (cr), Thomas Kennedy; 13 (tr), Carlyn Iverson; 14 (bl), David Merrell/Suzanne Craig Represents Inc.; 16 (tr), David Merrell/Suzanne Craig Represents Inc.; 17 (cr), David Kelley; 18 (tr), Thomas Kennedy; 22 (tr), David Kelley; 23 (bl), Thomas Kennedy; 25 (cr), Thomas Kennedy; 27 (tr), David Kelley; 30 (br), David Kelley; 34 (tr), Thomas Kennedy; 36 (br), David Kelley; 41 (bl), Thomas Kennedy; 45 (cr), Thomas Kennedy; 53 (tr), Layne Lundstrom; 55 (bl), Accurate Art, Inc.; 56 (tr), David Merrell/Suzanne Craig Represents Inc.; 57 (bl), David Kelley; 65 (tr), Thomas Kennedy; 67 (bl), Thomas Kennedy; 68 (b), David Kelley; 69 (tr), David Kelley; 70 (tr), Thomas Kennedy; 72 (tr), Thomas Kennedy; 74 (c), David Kelley; 78 (tr), Dodson Publications Services; 80 (tr), David Kelley; 82 (bl), Dodson Publications Services; 83 (tr), Thomas Kennedy; 86 (tr), Accurate Art, Inc.; 86 (br), Accurate Art, Inc.; 88 (bl), Thomas Kennedy; 90 (bl), David Kelley; 92 (b), David Kelley

Printed in the United States of America

ISBN 0-03-054417-3

10   085   05 04 03

• CONTENTS •

PHYSICAL SCIENCE DEMONSTRATIONS

CONTENTS, CONTINUED

CONTENTS, CONTINUED

Safety Guidelines and Symbols

As well as being great motivators, demonstrations are also often safer alternatives to student lab activities. It is important to keep in mind basic safety information such as the following, however, to guarantee your safety and the safety of your student audience.

- **GENERAL** Always try a demonstration with an adult partner before performing it for the class. Make sure students are a safe distance from the demonstration site. Share with the class the location of all safety equipment and the proper procedure for using them. Make sure students know the school's fire-evacuation routes. Keep the demonstration area free from unnecessary clutter.

EYE SAFETY Wear approved safety goggles when working with any chemicals or mechanical devices, or near any type of flame or heating device.

HAND SAFETY Avoid chemical or heat injuries to your hands by wearing protective gloves or oven mitts. Check the materials list in the lab for the type of hand protection you should wear while performing the experiment.

CLOTHING PROTECTION Wear an apron to protect your clothing from staining, burning, or corrosion.

SHARP/POINTED OBJECTS Use extreme care when handling knives and other sharp instruments.

HEAT Wear approved safety goggles when using a heating device or a flame. Wear oven mitts to avoid burns.

ELECTRICITY Do not use equipment with damaged cords. Be careful where equipment and cords are placed to avoid accidents. Ensure that your hands are dry and that electrical equipment is turned off before you plug it into an outlet.

CHEMICALS Wear approved safety goggles when handling chemicals. Read chemical labels. Wear an apron and latex gloves when working with acids or bases. If a spill gets on your skin or clothing, rinse it off immediately with water for at least 5 minutes.

ANIMAL AND PLANT SAFETY Wash your hands thoroughly after handling an animal or any part of a plant.

GLASSWARE Examine all glassware for chips and cracks before using it. Use plastic substitutes and heat-resistant glass whenever possible.

CLEANUP Wash glass containers with soap and water. Put away all equipment and supplies. Properly dispose of all chemicals and other materials. Make sure water, gas, burners, and hot plates are turned off. Make sure all electrical equipment is unplugged. Wash hands with soap and water after performing a demonstration.

The Scientific Method

The steps that scientists use to answer questions and solve problems are often called the scientific method. The scientific method is not a rigid procedure. Scientists may use all of the steps or just some of the steps. They may even repeat some steps. The goal of a scientific method is to come up with reliable answers and solutions.

Six Steps of a Scientific Method

1. Ask a Question Good questions come from careful **observations.** You make observations by using your senses to gather information. Sometimes you may use instruments, such as microscopes and telescopes, to extend the range of your senses. As you observe the natural world, you will discover that you have many more questions than answers. These questions drive the scientific method.

Questions beginning with *what, why, how,* and *when* are very important in focusing an investigation, and they often lead to a hypothesis. (You will learn what a hypothesis is in the next step.) Here is an example of a question that could lead to further investigation.

Question: How does acid rain affect plant growth?

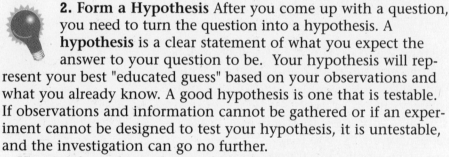

2. Form a Hypothesis After you come up with a question, you need to turn the question into a hypothesis. A **hypothesis** is a clear statement of what you expect the answer to your question to be. Your hypothesis will represent your best "educated guess" based on your observations and what you already know. A good hypothesis is one that is testable. If observations and information cannot be gathered or if an experiment cannot be designed to test your hypothesis, it is untestable, and the investigation can go no further.

Here is a hypothesis that could be formed from the question, "How does acid rain affect plant growth?"

Hypothesis: Acid rain causes plants to grow more slowly.

Notice that the hypothesis provides some specifics that lead to methods of testing. The hypothesis can also lead to predictions. A **prediction** is what you think will be the outcome of your experiment or data collection. Predictions are usually stated in an "if...then" format. For example, if meat is kept at room temperature, then it will spoil faster than meat kept in the refrigerator. More than one prediction can be made for a single hypothesis.

Here is a sample prediction for the acid rain hypothesis.

Prediction: If a plant is watered only with acid rain (which has a pH of 4), then the plant will grow at one-half its normal rate.

3. Test the Hypothesis After you have formed a hypothesis and made a prediction, it is time to test your hypothesis. There are different ways to test a hypothesis. Perhaps the most familiar way is by conducting a controlled experiment. A **controlled experiment** is an experiment that tests only one factor at a time. A controlled experiment has a **control group** and one or more experimental groups. All the factors for the control and **experimental groups** are the same except for one factor, which is called the **variable.** By changing only one factor (the variable), you can see the results of just that one change.

Sometimes, a controlled experiment is not possible due to the nature of the investigation. For example, stars are too far away, dinosaurs have been extinct for millions of years, and the Earth's core is surrounded by thousands of meters of rock. It would be difficult if not impossible to do controlled experiments on such things. Under these and many other circumstances, a hypothesis may be tested by making detailed observations. Taking measurements is one way of making observations.

4. Analyze the Results After you have completed your experiments, made your observations, and collected your data, you must analyze all the information you have gathered. Tables and graphs are often used in this step to organize the data.

5. Draw Conclusions Based on the analysis of your data, you should conclude whether your results support your hypothesis. If your hypothesis is supported, you (or others) might want to repeat the observations or experiments to verify your results. If your hypothesis is not supported by the data, you may have to check your procedure for errors. You may even have to reject your hypothesis and make a new one. If you cannot draw a conclusion from your results, you may have to try the investigation again or carry out further observations or experiments.

6. Communicate Results After any scientific investigation, you should report your results. By doing a written or oral report, you let others know what you have learned. They may want to repeat your investigation to see if they get the same results. Your report may even lead to another question, which in turn may lead to another investigation.

Air Ball

Purpose

Students are presented with a tricky problem to solve in order to prepare them for a discussion about the scientific method.

Time Required

10–15 minutes

Lab Ratings

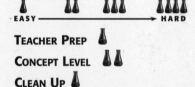

EASY ——————————→ HARD

TEACHER PREP

CONCEPT LEVEL

CLEAN UP

MATERIALS

- masking tape
- 2 plastic cups (about 250 mL or 9 oz)
- metric ruler
- table-tennis ball

What to Do

1. Securely tape each cup to a table so that the cups are about 3 cm apart.

2. Place a table-tennis ball in one of the cups.

3. **Ask a Question** Present the following challenge to students: How can you move the ball from one cup to the other without touching the ball, touching the cups, or using any tools or materials?

4. **Form a Hypothesis** Ask students: Does anyone have any ideas about how this could be done? Record students' ideas under the heading Hypotheses on the blackboard.

5. **Conduct an Experiment** After you have recorded several hypotheses, ask the class to determine which one seems the most feasible. Then call on volunteers to test that hypothesis.

6. **Analyze the Results** If the experiment isn't successful, ask for feedback from the students. Explain to students that trial and error is an important part of the scientific method. Add new hypotheses or revise existing hypotheses as necessary. You may also wish to call on a student to make notes on the blackboard about why each hypothesis that fails does so.

7. **Draw Conclusions** Call on new volunteers to test a new hypothesis. Continue this process until a solution is reached. *(Answer: Blow a short, hard gust of air at an angle toward the ball. This will push the ball from one cup to the next. Students will probably need to try several angles before they succeed.)*

Explanation

When you blow toward the ball at an angle, some of the air enters the cup and some of the air flows across the top of the cup. The air entering the cup sweeps the ball out of the cup. According to Bernoulli's principle, pressure within the air current moving over the top of the cup is lower than the pressure inside the cup. The low pressure acts as a partial vacuum, which captures the rising ball and carries it into the second cup.

Point out to students that as they attempted to solve this problem, they used steps known as the scientific method. They began with a clear problem, formed a hypothesis, conducted an experiment, analyzed the results, and drew conclusions.

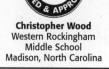

Christopher Wood
Western Rockingham
Middle School
Madison, North Carolina

DISCOVERY LAB

Getting to the Point

Purpose

Students use scientific methods to make sense of a discrepant event.

Time Required

10–15 minutes

Lab Ratings

EASY ————————————→ HARD

TEACHER PREP

CONCEPT LEVEL

CLEAN UP

MATERIALS

- metal nail file
- aluminum knitting needle (size 2 or 3), or a very large upholstery needle
- glycerin
- 5–6 tissues
- 12 in., round, helium-quality balloons (2)

Advance Preparation

You may wish to practice this demonstration before performing it in the classroom. Prepare the knitting needle by filing its pointed end to a very sharp point. Next use a tissue to apply a coat of glycerin to the needle. Prepare a few extra tissues with glycerin, and have them available in class.

What to Do

1. In front of the class, blow up a balloon and tie the end. Don't inflate the balloon all the way. There should still be room for the balloon to expand and a thicker portion near the tie. Repeat with the other balloon.

2. Show students the balloons and the needle, and ask them to predict what would happen if you were to stick the needle into the balloons. *(Expected answer: The balloons will pop.)*

3. Ask students: Do you mean like this? Then quickly insert the needle through the side of one balloon, as shown below. The balloon should pop loudly.

4. Then explain to students that you are now going to conduct some magic. With great fanfare, coat the needle with the tissue containing glycerin. Try to do so in a dramatic way that seems to serve no purpose other than being part of your "magic act."

5. Continue your magic routine by inserting the needle though the thick part at the bottom of the balloon with exaggerated care. Then dramatically push the needle all the way through the balloon and out the upper end of the balloon, as shown on the next page. The balloon should not pop.

continued...

Christopher Wood
Western Rockingham
Middle School
Madison, North Carolina

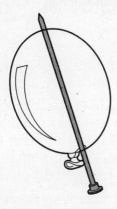

6. Pull the needle carefully out of the balloon. It should deflate slowly, revealing two neat holes.

Discussion

Ask if anyone can explain what happened. Students will probably comment on the different procedures that were used for each balloon. Ask students: Why would it make any difference in which direction I poked the balloon with the needle? Encourage students to examine the balloons up close and to offer an explanation. You may even wish to have students divide into small groups. Provide each group with materials, and encourage the students to try to repeat the experiment. Then encourage a class discussion about the successes and failures of each group.

Explanation

The inflated balloon is filled with air at a much higher pressure than the air around it. The reason the first balloon popped is that the needle created friction when it pierced the surface of the balloon. This friction caused a small tear, which the pressurized air inside the balloon rushed through. As the air sought to equalize the pressure, the balloon's surface began to tear more until all the air was released. The loud pop that you heard was caused by the violent collision of two air masses with different pressures. Similar to thunder, the colliding air created a shock wave that was transmitted by the air molecules to your ears.

You were able to pierce the second balloon without popping it because the material was thick and unstretched at the top and bottom. In addition, the glycerin that was applied to the needle reduced the friction enough to allow the needle to enter the balloon smoothly without tearing it. This is why the second balloon did not pop, even when you pulled the needle out.

Tell students that they will learn more about the science of friction and air pressure in later units. In the meantime, this activity is designed to help them understand that science is self-correcting. If one theory turns out to be wrong, it can be changed. It all begins with observations and predictions. Those predictions can be confirmed or denied through experimentation. If the predictions turn out to be wrong, it's time to develop more theories and conduct more experiments. This is what science is all about—changes over time.

Grand Strand

Purpose
Students gain perspective on the size of DNA strands as they watch you extract a long chain of DNA from the cells of a calf thymus.

Time Required
30–45 minutes

Lab Ratings

EASY ————→ HARD

TEACHER PREP

CONCEPT LEVEL

CLEAN UP

MATERIALS
- piece of calf thymus (called throat sweet-bread) about the size of a walnut
- sharp knife
- 100 mL of cold ethanol (95%)
- container of ice
- 10 mL of dishwashing liquid
- 1.5 g of NaCl (table salt)
- 90 mL of distilled water
- stirring rod
- 53 g of sugar
- 500 mL of water in a beaker
- protective gloves
- blender
- small funnel
- 4 pieces of cheesecloth
- 500 mL beaker
- inoculating loop or glass rod

Safety Information
Caution: Ethanol is extremely flammable and poisonous. Be sure to wear gloves and wash your hands after this activity and clean up any spilled liquid immediately.

Advance Preparation
- Purchase a calf thymus (throat sweet-bread) from a butcher. Cut it into small pieces about the size of a walnut, and freeze them until they are needed.
- Keep a bottle of ethanol in a freezer overnight. Make certain that the bottle is not completely full and that the cap is slightly loose. Keep the ethanol on ice during the class period.
- Prepare a detergent solution by combining 10 mL of dishwashing liquid with 1.5 g of NaCl (table salt) and enough distilled water to produce 100 mL of solution.
- Prepare an isotonic sucrose solution by adding 53 g of sugar to the beaker of 500 mL of water. Stir until all of the sugar is dissolved. Leave this solution on ice during the experiment.
- Wear gloves, and avoid touching the end of the glass rod or inoculating loop and the inside of the containers. The enzymes from your skin could fragment the DNA so that it will not form the long chains. To further protect against enzymes, rinse the glassware in 95 percent ethanol.

Introduction
A eukaryotic cell stores its DNA within a specialized structure called the nucleus. DNA is the genetic material for the organism. This molecule determines what the structure and function of cell proteins should be. Each cell contains all of the genetic information for the entire individual.

DNA is one of the longest molecules in the cell. Because a strand of DNA contains so much information, it is too long to fit inside the cell without curling and condensing. Tell students that in this experiment, you will remove DNA from a cell and demonstrate the DNA's incredible size.

continued...

Kenneth Creese
White Mountain Jr. High
Rock Spring, Wyoming

What to Do

1. Add a small piece of thymus (about the size of a large walnut) to the blender. Pour about two-thirds of the sucrose solution into the blender, and blend the ingredients until the mixture is the consistency of a milkshake. If necessary, add more of the sucrose solution and blend the mixture to achieve the correct consistency. Explain to students that the action of the blender has broken the cell membranes and has left a mixture of cytoplasm, organelles, and sucrose solution.

2. Next filter large cell debris, such as membrane fragments. Line the funnel with four layers of cheesecloth, and place it over the 500 mL beaker. Pour the thymus blend through the funnel. Discard the cheesecloth and its contents.

3. You have now collected a cell suspension in the beaker. You should have approximately 100 mL of the suspension. If you have more than 100 mL, discard the excess suspension.

4. Tell students that you will now break down the nuclear membranes. Add the detergent solution to the cell suspension so that you have a 1:1 ratio. In other words, if you have only 95 mL of the cell suspension, then add 95 mL of the detergent solution.

5. Shake the beaker GENTLY from side to side—don't let the mixture get foamy.

6. To separate the DNA from the nuclei, SLOWLY pour the ice-cold ethanol into the beaker with the nuclei-detergent mixture. Pour it carefully along the side of the beaker so that a layer of ethanol forms on top of the mixture.

> **SAFETY ALERT!**
>
> Ethanol is a hazardous substance. It is extremely flammable and poisonous. Be sure to wear protective gloves and clean up any spilled liquid immediately.

7. Watch the precipitate form. The two liquids will form two layers. The precipitate will form where these two layers meet. The precipitate is a long chain of DNA molecules. Note: If the precipitate does not form immediately, put the mixture on ice until the precipitate forms.

8. Use your inoculating loop or glass rod to gently stir the precipitate layer. Try to keep the loop or rod out of the lower layer of liquid.

9. Gently pull out the long chains of DNA. You may need to twirl the glass rod slightly to extract the chains. These DNA strands can be up to 2 m long! Explain to students that each strand consists of strands of DNA from several nuclei.

Discussion

Students may be interested to know that DNA is one of the longest molecules in the body. If all of the DNA in a human body were stretched out, it would be 300 million kilometers long. That is the distance from the Earth to the Sun and back! Why is DNA such a long molecule? *(Each cell in an organism contains all of the genetic material for the entire organism— that's a lot of coding!)*

LIFE SCIENCE

MAKING MODELS

Stop Picking on My Enzyme

Purpose

This activity is a fun, simple way to familiarize students with the vocabulary and processes of enzyme action.

Time Required

10–15 minutes

Lab Ratings

EASY ——————————→ HARD

TEACHER PREP

CONCEPT LEVEL

CLEAN UP

MATERIALS
• 2 flat-ended toothpicks

Introduction

Enzymes are proteins used by cells to stimulate and direct chemical processes. Without the assistance of an enzyme, many of the cellular processes in the body would occur much more slowly or would not occur at all.

Enzymes help chemical reactions occur more efficiently by decreasing the *activation energy,* or the energy required for a reaction to begin. The molecules that react with the enzyme are called substrates. Some enzymes work by pulling two substrates close to each other. Other enzymes work by weakening the chemical bonds between two atoms within a molecule.

When the pH or temperature within a cell or the bloodstream changes, enzymes are denatured. This means that their shape changes and they are no longer able to work properly.

HELPFUL HINT
You can easily involve students in this demonstration by giving them toothpicks and asking them to follow the steps along with you.

What to Do

1. Ask students to imagine that your hand is an enzyme. Pinch your thumb and forefinger together, and tell students that the area between the tips of your fingers is the enzyme's active site.

2. Pick up a toothpick with one hand, and hold it in the "active site." Explain that the toothpick now represents the substrate. Break the toothpick. The break in the toothpick represents the place where the chemical bonds have been weakened by the enzyme and subsequently broken.

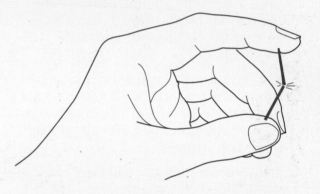

3. Tell students that enzymes are substrate-specific. This means that the shape of an enzyme's active site determines which substrate will bond to it. If something acts on the protein to change its shape, the enzyme can no longer work. For example, if an enzyme's environment becomes too warm or too acidic, the protein's shape will become denatured.

continued...

Elizabeth Rustad
Crane Jr. High
Yuma, Arizona

4. Reshape your hands by either bending or crossing your fingers. Explain that now the pH surrounding the enzymes has changed drastically. Try again to pick up and break a toothpick. Students should see that it is more difficult to break the toothpick when your hands are in this position.

USEFUL TERMS

activation energy
the energy required to begin a chemical reaction

enzyme
proteins used to lower the activation energy of chemical reactions in the body

substrate
molecule on which an enzyme acts

activation site
pocket formed in the folds of an enzyme; the reaction catalyzed by the enzyme occurs at the active site

---

Discussion

Use the following questions as a guide to encourage class discussion:

• How were my bent fingers similar to a denatured enzyme? *(Expected answer:*

When your fingers were bent, you could not pick up or break the toothpick. In much the same way, a misshapen enzyme is unable to bond to its substrate or to catalyze a reaction.)

• Enzymes are not changed by the chemical reactions they catalyze. Instead, they lower the energy required to begin a reaction. How did the enzyme modeled in this demonstration lower the activation energy? *(This model enzyme lowered the activation energy by weakening the chemical bonds to make it easier to break down the molecule.)*

• **Critical Thinking** What are some examples of the function of enzymes in the body? *(Sample answer: Some enzymes are used to make carbohydrates and lipids; others are used to break down carbohydrates and lipids. Other enzymes are used to make other proteins.)*

LIFE SCIENCE

MAKING MODELS

It's in the Bag!

Purpose
Students are challenged to simulate the process of endocytosis in cells.

Time Required
10–15 minutes

Lab Ratings

EASY ———————→ HARD

TEACHER PREP 🧪
CONCEPT LEVEL 🧪🧪
CLEAN UP 🧪

MATERIALS
- clear-plastic shopping bag
- 2–4 pieces of candy
- 15 cm of string or several twist ties
- scissors

What to Do
1. Remind students that a cell is a membrane filled with fluid, much like a water balloon. Ask students how cells ingest large particles. Do they have mouths? Explain that the process of endocytosis allows cells to ingest particles without bursting.

2. Tell students that you are going to model endocytosis using the assembled materials. The bag represents the cell, and the candy represents food that the cell takes in. Tell them that your challenge is to bring the candy into the bag without making a hole in the bag while keeping your hands inside the bag at all times. Ask students to brainstorm about how this could be done.

3. Arrange the candy in a cluster on the desktop, and put the string and the scissors inside the bag. Keep your hands inside the bag. From this point forward, your hands will remain in the bag.

4. Cut the string in half. With your hand still inside the bag, close your hand around the candy. This should form a pouch, enclosing the candy within the bag. Tightly tie off the neck of the pouch with one string.

5. Use the second piece of string to tie off the neck of the pouch about 2 cm away from the first string. Use the scissors to cut the bag between the strings. Voila!

Explanation
Endocytosis is the cellular process in which the cell engulfs and imports large substances. During endocytosis, the cell membrane surrounds large food particles that are brought inside by membrane-bound pouches called *vesicles*. The inside of the cell is never exposed to the external environment.

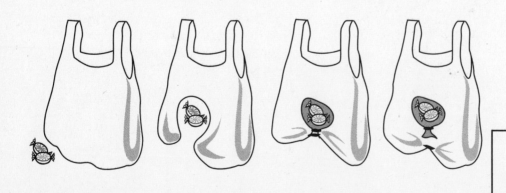

Elizabeth Rustad
Crane Jr. High
Yuma, Arizona

MAKING MODELS

Adaptation Behooves You

LIFE SCIENCE ▲ ▲ ▲

Purpose

Students explore how an animal's adaptations affect its chances for survival.

Time Required

15–20 minutes

Lab Ratings

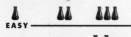

EASY ————————→ HARD

TEACHER PREP 🧪🧪

CONCEPT LEVEL 🧪🧪🧪

CLEAN UP 🧪

MATERIALS

- variety of objects to simulate a rocky terrain, such as books, staplers, and small rocks
- 2 small plastic sandwich bags
- 250 mL of rice
- small paper cup
- metric ruler

What to Do

1. Scatter several objects over a desktop.

2. Fill both plastic bags with 125 mL of rice and seal them. Place one bag in the paper cup.

3. Tell students that the objects on the desktop represent a mountainous terrain and that the bags of rice represent different hooves of two goats of the same species. The rice in the cup represents an individual with hard hooves, and the bag of rice alone represents an individual with soft, padded feet. Explain that this demonstration will simulate how adaptation and evolution occur within a population.

4. Ask students which hoof they believe is better adapted to the rugged terrain. *(Accept all reasonable responses.)*

5. Call on several volunteers. Ask one of the students to drop the cup with the rice from 30 cm above the "terrain." Repeat this 10 times, asking the other volunteers to assist. On the board, record the number of times the "hoof" lands securely on the ground.

6. Repeat this experiment using the bag of rice. Ask students which model hoof performed better. *(Expected answer: The soft hoof landed more solidly on the "terrain.")*

7. Were your hypotheses correct? *(Sample answer: My hypothesis was wrong. I had expected the hard hoof to be better adapted to the rough terrain.)*

Discussion

Tell students that mountain goats feed on grass, and ask them to suppose that grass is plentiful at the base of a mountain. The land is much flatter at the base, but the area is a common hunting ground for wolves that prey on mountain goats. Use the following questions as a guide to encourage class discussion:

- What might be an advantage and a disadvantage of soft hooves? *(Soft hooves may aid an escape by making climbing easier, but they would impede running on flat ground.)*

- What might be an advantage and a disadvantage of hard hooves for goats? *(Hard hooves may enable goats to run faster on flat ground, but they would make climbing difficult.)*

continued...

CLASSROOM TESTED & APPROVED

Kevin A. Tierney
Rolling Hills Middle School
El Dorado, California

- **Critical Thinking** Which goat is more likely to live longer and pass on its genes? *(It depends on the environment the goats live in.)*

Explanation

The process by which an advantageous genetic feature increases in a population is called *natural selection.* According to this theory, individuals best adapted to their environment are more likely to become adults and pass on their genes. Stress to students that genetic mutations occur all the time and that many have no effect on an individual's chances for survival. Genetic mutations occur randomly; they are not a direct response to an environmental condition.

The process of natural selection ensures that the individuals best suited to the specific environment survive. It takes many generations to eliminate one gene or to select another. Remind students that a mutation is not a negative change; it is merely a change. The environment does not determine whether a mutation occurs, but whether the animal with the mutation thrives.

DISCOVERY LAB

Unleash the Yeast!

LIFE SCIENCE

Purpose

Students observe as yeast convert food to energy, and learn that fungi are heterotrophic.

Time Required

15–20 minutes

Lab Ratings

EASY —————→ HARD

TEACHER PREP

CONCEPT LEVEL

CLEAN UP

MATERIALS

- 125 mL of warm water
- 60 mL of sugar
- 355 mL tall glass bottle
- envelope of active dry yeast
- balloon
- watch or a clock that indicates seconds
- paper towels

Advance Preparation

All of the materials and measurements are simply recommendations. You may wish to try this activity in advance and experiment with the materials so that you get the most dramatic effect within the desired amount of time. Keep the yeast packet refrigerated before the experiment to ensure that you have a live, active culture. Also stretch the balloon before the experiment so that it will be easier to inflate.

What to Do

1. Combine the sugar and warm water in the bottle, and shake the bottle to mix the solution. Add the yeast.

2. Quickly place the balloon over the neck of the bottle, and agitate the bottle for a few seconds. Wait at least five minutes.

3. Ask students to describe what is happening. *(The liquid is bubbling continuously, and the balloon is inflating slightly.)*

HELPFUL HINT

Be careful while removing the balloon from the bottle because there may be a foamy, gooey substance inside the balloon. Have paper towels on hand.

Discussion

Use the following question as a guide to encourage class discussion:

- Why did the balloon inflate? *(As a by-product of the digestion process, the yeast produced enough gas to inflate the balloon. As the yeast digested the sugar, alcohol and carbon dioxide were created as waste products. These products caused the liquid to bubble and the balloon to inflate.)*

- **Critical Thinking** Yeast is a fungus, which is a different category than an animal or a plant. Is the yeast more like a plant or an animal in the way that it obtains its food? Explain. *(The yeast has no chlorophyll and cannot produce its own food like plants. Therefore, it is more like an animal in the way it obtains its food.)*

Explanation

Fungi, such as yeast, are *heterotrophic*. This means that they do not produce their own food. Animals are also heterotrophic. They need to hunt or gather their food in order to survive. Plants, on the other hand, produce their own food. Plants use sunlight to produce the energy they need to live. Organisms that produce their own food, such as plants, are called *autotrophs*.

Kathy LaRoe
Radley Elementary School
East Helena, Montana

MAKING MODELS

Inner Life of a Leaf

Purpose

This activity demonstrates how the vascular system of a plant functions and introduces students to plant structures.

Time Required

10–15 minutes

Lab Ratings

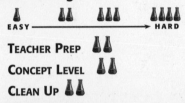

EASY ——————→ HARD

TEACHER PREP
CONCEPT LEVEL
CLEAN UP

MATERIALS

- soda bottle
- tap water
- ivy plant
- modeling clay about the size of a walnut
- plastic drinking straw
- scissors

What to Do

1. Fill a soda bottle with water to within 3 cm of the top.

2. Cut a leaf from the ivy plant so that the stem is approximately 8 cm. Wrap a piece of clay around the stem, near the leaf.

3. Place the stem into the bottle so that its end is below the surface of the water. Cover the opening of the bottle with the clay, making sure that the bottle is tightly sealed.

4. Use a pencil to poke a hole through the clay to make an opening for the straw.

5. Position the bottom of the straw just inside the mouth of the bottle, above the water. Seal the clay around the straw, and double-check that the rest of the bottle opening is still sealed.

6. Ask students to watch the plant carefully as you inhale air from the bottle. (Be careful to keep the straw above the water!) Since the clay has created a tight seal, withdrawing air from the bottle will be a challenge.

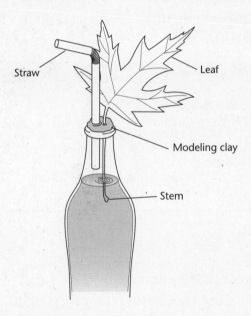

Straw — Leaf

Modeling clay

Stem

Explanation

Holes called stomata perforate the leaf. Tiny tubes called xylem and phloem run up and down the stem. The *xylem* carries water up from the roots to the stem and leaves just like the straw carried air to your mouth. The *phloem* carries carbohydrates, which are produced by photosynthesis in the leaves

continued...

Lee Yassinski
Sun Valley Middle School
Sun Valley, California

and stems, to parts of the plant that need carbohydrates, like root tips and buds. The *stomata* are pores that allow the exchange of gases with the outside. Water vapor leaves the stomata as carbon dioxide enters. However, water vapor does not enter the plant through the stomata. Water enters from the roots.

Discussion

Use the following questions as a guide to encourage class discussion:

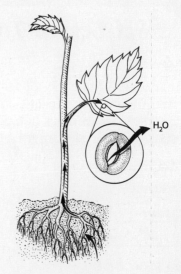

- Now that you know something about leaf structure, what do you think caused the bubbles to form at the bottom of the stem? *(As air was sucked out through the straw, more air was drawn into the bottle through the stomata and xylem. Because air was forced out of the bottle, the air pressure inside the bottle decreased. The leaf drew in air to maintain equilibrium.)*

- How did the leaf perform like the straw? *(The xylem carried water up from the roots to the stem and leaves, just as you drew air in through the straw.)*

LIFE SCIENCE

Six-Legged Thermometer

Purpose

Students observe the effect of temperature on the frequency of a cricket's chirping and learn how an animal's behavior may be affected by its environment.

Time Required

10–15 minutes

Lab Ratings

EASY ──────────────► HARD

TEACHER PREP
CONCEPT LEVEL
CLEAN UP

MATERIALS

- 2 mature male crickets (available in many bait and pet stores)
- 2 glass jars
- 2 nylon stockings
- 2 rubber bands
- refrigerator
- watch or a clock that indicates seconds
- Fahrenheit thermometer

Advance Preparation

Only the mature male crickets will chirp. Keep each cricket in a separate container; male crickets are extremely territorial and have a tendency to kill other males.

You may wish to try this activity in advance. First put one cricket in each jar. Cover each jar with a nylon stocking, and secure the stocking with a rubber band. (Be sure the stocking is stretched enough to allow sufficient oxygen

into the jars.) Leave one jar at room temperature. Leave the other in the refrigerator long enough to slow the chirping of the cricket significantly but not so long that the cricket is harmed. Note: Conduct this activity as soon as you remove the jar from the refrigerator. Record the temperature of the refrigerator in degrees Fahrenheit.

What to Do

1. Have students observe the cricket in the refrigerated jar. Ask them to count the number of chirps the cricket makes in 15 seconds and then to add 40 to that number. Record both numbers on the board.

2. Record the temperature of the refrigerator in degrees Fahrenheit on the board. Point out the similarity between this number and the data from step 1; they should be roughly equal.

3. Have students repeat step 1 with the other cricket.

4. Measure and record the temperature of the room in degrees Fahrenheit. Again, the temperature of the room and the data from step 3 should be almost equal.

Discussion

Use the following questions as a guide to encourage class discussion:

- What can you conclude from your observations? *(Sample answer: There is a relationship between temperature and chirping frequency. A cricket can serve as a kind of thermometer.)*

- If we refer to the number of chirps in 15 seconds as degrees cricket, what is the corresponding temperature in degrees cricket for 82° F? *(82 − 40 = 42 degrees cricket)*

Kevin A. Tierney
Rolling Hills Middle School
El Dorado, California

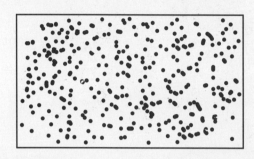

DISCOVERY LAB

The Fish in the Abyss

Purpose

This demonstration introduces students to the role of camouflage, or protective coloration.

Time Required

10–15 minutes

Lab Ratings

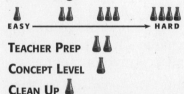

EASY ———————→ HARD

TEACHER PREP

CONCEPT LEVEL

CLEAN UP

MATERIALS

- 2 sheets of construction paper, both the same dark color
- scissors
- correction fluid or crayon
- plastic transparency sheet

Advance Preparation

- Cut a fish shape from one piece of construction paper. Leave a handle protruding from the bottom.

- Using the correction fluid or crayon, mark a random pattern of dots on the fish and on the second piece of paper.

- Cover the fish and the paper with the transparency sheet. The plastic should hold the edges of the cutout flat against the paper background.

- Hold the setup in the front of the classroom, or tape the background and transparency to the wall or blackboard. Be sure that students will not be able to see the edges.

What to Do

1. Challenge students to distinguish the animal from the background. Ask them if they know what kind of animal it is.

2. It should be very difficult for students to see the fish. Tell them that the animal is well-camouflaged to protect itself from predators.

3. Move the fish under the plastic. Students should be able to see it more clearly.

4. Ask students for examples of animals that use camouflage for defense. Discuss different patterns and colors used for camouflage.

Explanation

Many fish and other animals have coloring that protects them from detection. The yellow perch, for example, is camouflaged by underwater vegetation. It is difficult to detect these fish when they are still. When they move, it is easier to distinguish them from the background because humans and many other animals use specialized brain cells to perceive motion. Information is relayed to these cells from light-sensitive cells at the back of the eye.

CLASSROOM TESTED & APPROVED

James Chin
Frank A. Day Middle School
Newtonville, Massachusetts

▶ LIFE SCIENCE

Voracious Fly Catcher

Purpose
Students learn about a carnivorous plant and discuss ways that it is specifically adapted to its environment.

Time Required
10–15 minutes

Lab Ratings
EASY ————————→ HARD

TEACHER PREP
CONCEPT LEVEL
CLEAN UP

MATERIALS

- Venus' flytrap (or other carnivorous plant)*
- small insect, such as a fly or spider
- tweezers
- * Be sure that the plant was grown in a greenhouse and not in the wild.

What to Do
Use a Venus' flytrap to introduce students to the subject of soil nutrients. Ask students if they recognize the plant. Although some students may be unfamiliar with the appearance of a Venus' flytrap, it is likely that most of the class has heard of the plant.

Ask students if they can name the unusual characteristic for which the plant is most well known. *(It eats insects.)*

If the plant has its leaves open, you may be able to demonstrate this unusual characteristic. Use the tweezers to place an insect in an open leaf. Students may be surprised to see the two lobes of the leaf shut, trapping the insect inside. Explain that the insect touched several sensitive hairs that exist on each lobe, causing the two halves to close.

Ask students if they have any idea what will happen to the insect. *(Expected re-*

sponse: The plant will eat it!) Ask students how a plant that has no digestive system or muscular tissue can eat an insect. *(The Venus' flytrap secretes a fluid through special glands in its leaves. This fluid digests the soft parts of the insect.)* Explain that because of this characteristic, the Venus' flytrap is considered a carnivorous, or meat-eating, plant.

Then ask the class, "Why is this plant an insect-eater?" Ask the class to brainstorm hypotheses. Explain that the answer relates to the kind of soil in the plants' native habitat. Tell students that the rare plant grows naturally in only a few locations in North Carolina and South Carolina. These locations are in a region where small swamps and open pine woodlands intersect. The soil in such locations is sandy and generally lacking in nitrogen. Many generations ago, a mutation occurred within the genes of an ancestor of Venus' flytrap that allowed it to extract nitrogen from insects rather than from the soil. Though the soil in these regions of North Carolina and South Carolina is nitrogen-deficient, the individuals with this mutation could thrive and reproduce there.

Tell students that most plants aren't carnivorous, so they must obtain their nitrogen and other important nutrients directly from the soil. We can help plants get the nutrients they need by composting or by applying fertilizers that consist of different combinations of soil nutrients. By doing so, we can grow plants in soil that might not otherwise be suitable for the plants.

Georgiann Delgadillo
Continuous Curriculum School
Spokane, Washington

Get the Beat!

LIFE SCIENCE

Purpose

By observing the vibration of a coffee stirrer placed on a pulse point, students are introduced to the circulatory system.

Time Required

10–15 minutes

Lab Ratings

EASY ——————→ HARD

TEACHER PREP
CONCEPT LEVEL
CLEAN UP

MATERIALS

- 1 plastic coffee stirrer per student
- modeling clay
- clock or watch that indicates seconds

HELPFUL HINT

The coffee stirrer may be replaced with a broomstraw or similar object. It is easiest to see the movement with objects 10 cm in length or longer.

What to Do

1. Give each student a coffee stirrer and a very small piece of clay. Ask students to attach the clay to one end of the stirrer.

2. Have students place one forearm on the desktop, palm side up. Ask them to locate the pulse point on their wrists.

3. Direct students to press the clay on their pulse point so that the stirrer is in a vertical position.

4. When the stirrer is in the correct position, it will slowly vibrate back and forth. Students may need to reposition the clay to be certain that it is on their pulse point.

5. Ask students to count the vibrations for one minute and record their results. Have them remove the coffee stirrer and set it aside.

6. Divide the class into two groups. Ask the first group to run in place for two minutes and the other group to sit with their heads resting on the table for two minutes. Students should then replace the stirrer and measure their pulse again. Compare the results of the two trials.

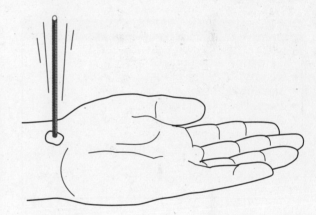

Explanation

The stirrer vibrates back and forth at a regular rate. This rate is the individual's pulse rate. Adults will have a pulse rate between 60 and 80 beats per minute. A child's pulse rate will be slightly higher, between 80 and 140 beats per minute.

The heart contracts at a regular rate to send blood through the body. The contractions force blood through the blood vessels. All blood vessels throb as the heart contracts, but the vessels near the wrist are close enough to the surface of the skin to be felt. The pulse can be measured in several other places on the body, including the neck.

CLASSROOM TESTED & APPROVED

Georgiann Delgadillo
Continuous Curriculum School
Spokane, Washington

MAKING MODELS

Take a Deep Breath

Purpose

With this model of the respiratory system, students learn about the breathing process.

Time Required

15–20 minutes

Lab Ratings

EASY ———————→ HARD

TEACHER PREP

CONCEPT LEVEL

CLEAN UP

MATERIALS

- 2 L plastic soda bottle
- scissors
- masking tape
- 2 small sandwich bags
- glass Y-tube
- 2 rubber bands
- one-hole stopper to seal the mouth of the bottle
- a large, round balloon

Advance Preparation

- Use the scissors to remove the bottom of the soda bottle. The cut edge of the soda bottle should be smooth so that it does not tear the balloon. You may wish to tape the ragged edge of the bottle with the masking tape.

- Tightly attach the two small sandwich bags to the Y-tubing with rubber bands. Insert this apparatus in the bottle as shown in the illustration at right. Slide the rubber stopper over the end of the Y-tube. Press the stopper into the neck of the bottle.

- Cut the large balloon in half. Stretch it across the bottom of the bottle. Secure

the balloon with glue or tape. The entire bottle should be airtight except for the end of the Y-tube which is exposed to outside air.

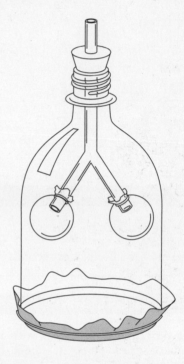

What to Do

1. Ask students to take a deep breath and to notice how their body responds. They should find that as they inhale, their chest enlarges, and as they exhale, their chest compresses.

2. Now show students the apparatus. Tell them that the big balloon represents the diaphragm. This is a muscle that expands and contracts to help animals breathe.

3. Encourage students to find their own diaphragm, explaining that the diaphragm lies between the rib cage and the abdomen.

continued...

CLASSROOM TESTED & APPROVED

Dennis Hanson
Big Bear Lake Middle School
Big Bear Lake, California

4. Tell students that the straight part of the Y-tube represents the trachea, which carries air from the nose and mouth. The trachea separates into two bronchi that channel air to the lungs. The lungs are represented by the sandwich bags. The bottle simulates the rib cage. You may wish to qualify this setup by telling students that the chest cavity in animals expands and contracts more than the bottle will because the rib joints in animals are flexible.

5. Now push on the balloon "diaphragm." The pressure inside the bottle increases as you push on the balloon. In response, the two sandwich bags contract and release air through the Y-tubing.

6. Pull on the center of the large balloon. As you pull down on the diaphragm, pressure inside the bottle decreases and the sandwich bags expand as they "inhale" air through the Y-tube.

Explanation

As you push on the balloon diaphragm, the pressure inside the bottle increases. The two sandwich bags contract and release air through the Y-tubing. As you pull down on the diaphragm, pressure inside the bottle decreases and the sandwich bags expand as they "inhale" air through the Y-tube.

Discussion

Use the following questions as a guide to encourage class discussion:

- What happens to the diaphragm, lungs, and chest cavity as a person inhales? *(As a person inhales, the diaphragm moves downward, causing the lungs and chest cavity to expand. This causes the lungs to suck air in through the bronchi.)*

- How did this demonstration model the body's actions during inhalation? *(As the balloon was pulled, pressure decreased inside the bottle. This expansion caused the sandwich bags to suck air in through the Y-tube, which represents the trachea.)*

- What happens to the diaphragm, lungs, and chest cavity as a person exhales? *(When a person exhales, the diaphragm moves upward, causing the lungs and chest cavity to compress. This pushes air out of the lungs and through the bronchi.)*

- After watching this, what did you learn about exhaling? *(As the large balloon was pushed, pressure inside the bottle increased. The sandwich bags then forced air out through the Y-tube in order to maintain equilibrium.)*

LIFE SCIENCE

Liver Let Live

Purpose

This simple, quick demonstration shows students how the small intestine breaks up oil and fats into tiny droplets.

Time Required

5–10 minutes

Lab Ratings

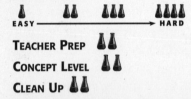

EASY —————→ HARD

TEACHER PREP
CONCEPT LEVEL
CLEAN UP

MATERIALS

- food coloring
- 500 mL jar with a lid
- 300 mL of tap water
- 100 mL of vegetable oil
- 10 g of bile salts or 3–4 drops of liquid detergent

What to Do

1. Add about 4 drops of food coloring to the water in the jar to make the water more visible.

2. Pour the oil into the jar. Tighten the lid, and shake the jar vigorously. Then let the jar stand for a few moments.

3. Ask students to describe what happened. *(The oil and water separated.)*

4. Remind students that many of the foods they eat may contain fats and oils. Tell them that you are going to simulate how the body begins to digest these oils when they reach the small intestine.

5. Add a few drops of liquid detergent or 10 g of bile salts to the suspension. Cover and shake the jar again.

6. Let the jar stand, and then ask students to observe the mixture. *(The oil has broken down into tiny droplets and combined with the water to produce a cloudy mixture.)*

Explanation

As you demonstrated when the jar was shaken, fats and oils are insoluble in water. When these materials reach the small intestine, they are digested with the help of an enzyme called lipase. However, before lipase can chemically digest the fats, the large globules must be physically broken into smaller droplets. A greenish liquid called bile, which is produced by the liver and stored in the gallbladder, helps break down the fat globules. Bile distributes broken down fat molecules evenly through the small intestine in a process known as emulsification. The cloudy mixture produced in this demonstration is similar to the emulsified liquid produced by bile.

Georgiann Delgadillo
Continuous Curriculum School
Spokane, Washington

DEMO

15 TEACHER-LED DEMONSTRATION

Now You See It, Now You Don't

Purpose
Students learn about persistence of vision.

Time Required
10–15 minutes

Lab Ratings
EASY ———————→ HARD

TEACHER PREP
CONCEPT LEVEL
CLEAN UP

MATERIALS

- slide projector, overhead projector, or video projector
- moveable projection screen, or large piece of paper
- 35 mm slide
- masking tape
- PVC pipe or back of wooden ruler (at least 3 cm in diameter and 1 m in length)

Advance Preparation
Use the projector to focus an image onto a movable screen or piece of paper in the classroom. The image should be in focus far enough away from the wall opposite the projector so that the picture is not identifiable on the wall. Use masking tape to mark the plane where the image is in focus. Remove the screen or paper, leaving the projector set up and adjusted.

What to Do
1. Turn on the projector. Hold the PVC pipe or ruler horizontally directly over the tape and at the level of the projected image so that students can see a small sliver of the slide focused on the pipe.

2. Ask students to predict whether they will be able to see the entire slide at one time on the PVC pipe.

3. Move the PVC pipe up and down through the plane of the focused image slowly so that students are not able to see the entire image.

4. Now move the PVC pipe up and down through the plane of the image very quickly so that students are able to see the entire image.

Explanation
The moving wand reflects the light just as the screen does, except that the wand reflects the image piece by piece. When this reflected light enters the eye, it makes an image on the retina. The eye retains each piece of the image for about 1/30 of a second, long enough to let the brain put the pieces together and form a composite picture. When the wand is moved slowly through the image, students are not able to see the picture because their eyes are not able to retain the image for a long enough period of time. When the wand is moved quickly, students see the entire image because the brain perceives the light to be constant.

The eye's tendency to retain an image for a fraction of a second is called *persistence of vision*. Rods and cones, the light detectors of the eye, continue to perceive the light even after it has come and gone.

Persistence of vision enables people to see images on a television screen or a computer monitor. These light sources flash the image on and off 60 or 120 times per second. Because the flashes are so quick, the rods and cones are constantly sending signals to the brain, and the image appears to be constant.

LIFE SCIENCE

Dennis Hanson
Big Bear Middle School
Big Bear Lake, California

Tubby Terra

Purpose
Students investigate why the Earth bulges at the equator.

Time Required
5–10 minutes

Lab Ratings

EASY ———————→ HARD

TEACHER PREP
CONCEPT LEVEL
CLEAN UP

MATERIALS

- metric ruler
- construction paper
- scissors
- glue
- paper-hole punch
- pencil

What to Do

1. Tell students that you will make a model of the Earth to demonstrate that it is not a perfect sphere.

2. Measure and cut two separate strips of construction paper that are each 3 × 40 cm.

3. Glue the strips together in an X shape. Bring the four ends of the X together at one point so that they overlap and form a sphere. Glue the ends in this position.

4. While you are waiting for the glue to dry, show students a globe, and discuss the shape of the Earth.

5. Punch a hole in the center of the over-lapped ends. Insert the pencil through the hole so that 5 cm of the pencil is in the center of the sphere, as shown in the diagram.

6. Hold the other end of the pencil between your palms, and spin the sphere by moving your hands back and forth.

Explanation

While the sphere was spinning, the strips flattened at the top and bottom and the center bulged. This paper sphere serves as a model to illustrate the effect the Earth's rotation has on its shape.

The revolving sphere demonstrates how centrifugal force acts on an object. Centrifugal force tends to pull a rotating object away from the center of rotation. More force is exerted on objects that are farther from the center. The Earth, like all rotating spheres, bulges at the middle and flattens at the poles. That is why the diameter of the Earth is 44 km greater at the equator than at the poles.

Dwight Patton
Carroll T. Welch Middle School
El Paso, Texas

MAKING MODELS

Settling Down

Purpose

Students learn about the formation of sedimentary rocks.

Time Required

5–10 minutes

Lab Ratings

EASY ——————→ HARD

TEACHER PREP

CONCEPT LEVEL

CLEAN UP

MATERIALS
• beaker or jar
• tap water
• sand (enough to fill $\frac{1}{8}$ of the beaker)
• pebbles (enough to fill $\frac{1}{8}$ of the beaker)
• soil (enough to fill $\frac{1}{8}$ of the beaker)
• stirring rod

What to Do

1. Tell students that in this demonstration they will learn about a process called sedimentation. To simulate sedimentation, you will use a beaker or jar of water to represent a river and a few common particles to represent sediments.

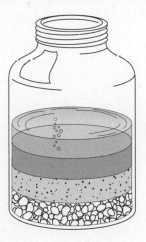

2. Fill the beaker or jar approximately halfway with water. Add a mixture of sand, pebbles, and soil to the water.

3. Stir vigorously, and then let the mixture stand until the particles settle.

Discussion

• Ask students to describe what happened. *(The particles in the mixture settled in layers, or became stratified, with the largest particles at the bottom.)*

• Ask students to explain how this demonstration simulates a river. *(As a river flows, it picks up and transports sediments. The heaviest particles settle to the bottom of the river, while smaller, lighter particles can be transported for greater distances.)*

• **Critical Thinking** What types of sediments are likely to be found at the point where a river meets an ocean? *(Fine particles, such as silt and clay, are found where a river meets an ocean.)*

• Explain that sediment can accumulate over thousands of years. The weight of the successive layers of sediment creates pressure on the deepest (oldest) layers. This pressure cements the sediments together, creating sedimentary rock.

Explanation

Particles that are transported in suspension by water, wind, or ice and later deposited are called sediments. Ultimately, layers of *sediments* harden and become *sedimentary rock.*

continued...

Dwight Patton
Carroll T. Welch Middle School
El Paso, Texas

EARTH SCIENCE

Generally, sediments that move by rolling or sliding downhill are deposited in a random mixture. However, sediments transported by wind or water are deposited according to size. Particles are dropped when the speed of the water or wind is too slow to carry them. Therefore, larger, heavier particles are deposited first. There are, of course, exceptions. Gold, for example, is dropped quickly on stream beds and beaches regardless of its size because it is unusually dense.

There are four major classes of particulate sediment: gravel, sand, silt, and clay.

A *delta* is the point where a river or stream meets a larger body of water. This area generally has a high amount of silt and clay because a river slows as it approaches a larger, slow-moving body of water. The larger particles are dropped well before the river reaches the delta. As sediment particles accumulate, their weight begins to compact the older sediments into rock. Water is forced away from the sediment, and chemicals cement the particles together over time.

18 **TEACHER-LED DEMONSTRATION**

Thar She Blows!

Purpose

Students see a working model of a geyser that erupts, resets itself, and erupts again!

Time Required

25–30 minutes

Lab Ratings

EASY ——————————→ HARD

TEACHER PREP

CONCEPT LEVEL

CLEAN UP

MATERIALS

- plastic food container
- sharp knife
- ring stand and metal ring
- tap water
- hot plate or other heat source
- glass tubing (.5–1 m long)
- Erlenmeyer flask
- one-hole rubber stopper
- plumber's putty about the size of a walnut

Safety Information

Be certain that students do not approach the hot plate. Students should stand back 2 m from the activity. No one should be allowed to stand over the glass tubing during the demonstration.

Advance Preparation

Drill a hole in the center base of your plastic food container. (A sharp, pointed knife twirled in a circle will work.) The hole should be just large enough for the glass tubing to be snugly inserted.

What to Do

1. Ask students if anyone has ever seen a geyser. If so, ask him or her to describe it. If not, explain that geysers are natural springs from which water or steam erupts at regular intervals. Old Faithful, in Yellowstone National Park, for example, is one of the world's most famous geysers. It erupts almost every hour for about 5 minutes.

2. Tell students to stand back—you are going to be building a model geyser!

3. Fill the Erlenmeyer flask to the rim with water.

4. Place the hot plate next to the ring stand, and set the flask on top of the hot plate.

5. Insert one end of the glass tubing into the top of a one-hole rubber stopper. Slide the glass tubing into the stopper until the tubing's bottom edge is flush with the bottom of the stopper.

6. Cap the Erlenmeyer flask with the assembled stopper and glass tubing.

7. Slide a metal ring onto the ring stand and over the glass tubing.

continued...

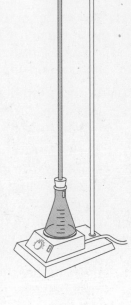

Dwight Patton
Carroll T. Welch Middle School
El Paso, Texas

WHIZ-BANG DEMONSTRATIONS **25**

Copyright © by Holt, Rinehart and Winston. All rights reserved.

8. Insert the other end of the glass tubing into the base of the plastic container. Adjust the glass tubing so that the top extends several centimeters above the bottom of the container but below the container's rim. To prevent leaking, seal the area where the glass tubing and container meet with plumber's putty.

9. Adjust the ring stand so that the ring is directly under the plastic container, and tighten the ring securely. Make sure that it is tight and secure enough to support the plastic container when it is filled with water.

10. Pour cool water into the plastic container until the glass tubing is 1–2 cm below the water's surface.

11. Ask students to predict what will happen when you turn on the heat. *(Expected answer: The water will boil, and steam will rise through the tube.)*

12. With the students standing back at least 2 m, turn on the hot plate and watch out!

13. As the hot water is released in an eruption, the geyser resets by drawing the cool water from the plastic container. The water in the plastic container will increase in temperature slightly with each eruption. The geyser will eject hot water several times before the temperature difference between the water in the two containers is not great enough to refill the system. At this point, you will have to turn off the heat source and allow the water to cool before repeating the procedure.

Explanation

The geyser created in this lab has the three features necessary for eruption: a heat source, a water supply, and a reservoir. The system worked because the heat source raised the temperature of the water in the Erlenmeyer flask above its boiling point. The superheated water vaporized into steam, rose through the glass tubing, and escaped out the top as an eruption. Once the pressure and heat were relieved, cool water from the plastic container replaced the water lost in the eruption. This lowered the temperature of the water in the tubing and flask, completing one eruption cycle.

Discussion

Use the following questions as a guide to encourage class discussion:

• Why do geysers erupt? *(Water trapped in the ground becomes heated and boils into steam, which forces its way out of the ground as an eruption.)*

• **Critical Thinking** What would the heat source be in a real geyser? *(Expected answer: Magma or geothermal energy)*

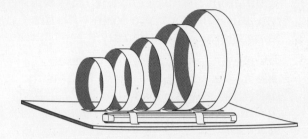

MAKING MODELS

When Buildings Boogie

Purpose

Students learn how different structures vibrate at different frequencies and learn why earthquakes affect neighboring buildings differently.

Time Required

10–15 minutes

Lab Ratings

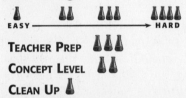

EASY ——————————→ HARD

TEACHER PREP
CONCEPT LEVEL
CLEAN UP

MATERIALS

- construction paper, about 20 × 50 cm
- metric ruler
- scissors
- masking tape or transparent tape
- sheet of cardboard, 30 × 8 cm
- ball bearing or BB
- plastic drinking straw

Advance Preparation

- Cut the construction paper into five strips 3 cm wide. The longest strip should be about 50 cm long, and each successive strip should be about 8 cm shorter than the previous one.

- Create rings by taping the ends of each strip together. Then tape the rings to the cardboard, arranging them in a row from largest to smallest, as shown in the diagram.

- Put the ball bearing or BB in the straw, and close both ends with a small piece of the construction paper. Secure the paper with tape. (Do not simply cover the ends with tape; you want the ball bearing to move freely in the straw and not become stuck to the tape.) The purpose of the straw is to hear the frequency.

- Attach the straw apparatus to the cardboard base with tape so that the straw is parallel to the rings. Move the cardboard back and forth a few times to be certain that you can hear the ball bearing tapping the ends of the straw. If you cannot, readjust the straw's position.

What to Do

1. Tell students that you are going to simulate an earthquake to see how different objects respond to the vibrations.

2. Begin to shake the cardboard base back and forth. Start very slowly, and continue to increase the frequency with which you move the cardboard back and forth. Explain to students that they can hear the frequency of the movement by listening to the ball bearing.

continued...

Vicky Farland
Crane Jr. High
Yuma, Arizona

EARTH SCIENCE

3. Encourage students to notice that different rings vibrate strongly, or *resonate*, at different frequencies. The rings will not all swing back and forth with the same rhythm and intensity.

4. Try moving the cardboard up and down. Ask students to observe how the pattern changes. *(They should notice that the rings resonate at different frequencies, depending on the direction that the base is shaken.)*

5. Students should notice that with both side-to-side and up-and-down movement, the largest ring will be the first to vibrate strongly, and the smallest will be the last.

6. After the smallest ring begins to vibrate, shake the cardboard faster. Notice that the largest ring will begin to vibrate strongly again. Each ring will vibrate strongly at more than one frequency. This additional frequency is a multiple of the ring's resonant or natural frequency.

Explanation

An object's stiffness and inertia determines at what frequency it vibrates most easily. Objects with stiffer structures and little inertia, like the smallest ring, tend to have higher resonant frequencies, while more-massive and elastic structures tend to have lower resonant frequencies.

During earthquakes, two buildings of different sizes may respond differently to the Earth's vibrations, depending on how closely their resonant frequencies match the earthquake's frequency. Of course, a building's design and the strength of the materials it is made of also contribute to whether a building is damaged by an earthquake.

Discussion

Use the following questions as a guide to encourage class discussion:

- Did the rings swing back and forth together? Why or why not? *(No; the largest ring began to vibrate first, and the smallest began to vibrate last. The largest ring was not as stiff as the smallest, so it had a lower resonant frequency.)*

- If the vibration of the cardboard simulates an earthquake, what do the rings represent? *(The rings represent buildings of different sizes.)*

- **Critical Thinking** Using what you learned in this activity, how do you expect different buildings to respond to an earthquake? *(Generally, taller buildings have lower resonant frequencies. They vibrate more easily and may be more damaged by a smaller earthquake than shorter, sturdier buildings.)*

MAKING MODELS

What Makes a Vent Event?

Purpose
Students witness a simulation of the explosion that occurs when hardened magma blocks the vent of a volcano.

Time Required
10–15 minutes

Lab Ratings

EASY ——————→ HARD

TEACHER PREP
CONCEPT LEVEL
CLEAN UP

MATERIALS

- paring knife
- potato or 2.25 cm cork (found in craft stores)
- 0.5 L plastic bottle with narrow top
- 150 mL of white vinegar
- 7.5 mL of baking soda
- 8 x 10 cm sheet of tissue paper or paper towel

Safety Information
Use caution when performing this demonstration. Make sure that the bottle is positioned so that the potato stopper will not strike anyone. Have students stand back at least 5 m.

What to Do

1. Tell students: Watch out—I am going to simulate an erupting volcano!

2. Use the knife to cut a stopper out of the potato. When complete, the stopper should fit tightly in the bottle opening and should be about 2 cm in length. Set the stopper aside until you are ready to use it.

3. Pour the vinegar into the bottle.

4. Wrap the baking soda into the tissue paper. Drop the tissue packet into the bottle.

5. Quickly insert the potato stopper into the bottle opening.

6. Step back, and watch the eruption!

Explanation
Huge pools of molten rock, known as magma, are located deep inside the Earth. A volcano forms when magma breaks through the Earth's surface. When a volcano erupts, the magma and other volcanic materials flow through the vent of a volcano. Magma may contain large amounts of trapped gases. These gases are sometimes released with explosive force during a volcanic eruption. Sometimes magma cools and hardens before it passes through the vent of the volcano. When this happens, more magma and gases are trapped inside the volcano. Great pressure builds beneath the blockage. This pressure ultimately forces the hardened magma away, and the volcano erupts.

Discussion
Use the following questions as a guide to encourage class discussion:

- Why did the explosion occurr? (*Gas forms when the vinegar and baking soda are mixed. The pressure of this gas pushes the potato stopper out of the bottle with a loud popping sound.*)

- What did the potato stopper represent in this simulation? (*The potato stopper represented hardened magma blocking the vent of a volcano.*)

CLASSROOM TESTED & APPROVED

Vicky Farland
Crane Jr. High
Yuma, Arizona

EARTH SCIENCE

MAKING MODELS

How's Your Lava Life?

Purpose

Students witness the flow of magma through a model volcano and observe the formation of dikes along which the magma flows.

Time Required

25–30 minutes

Lab Ratings

EASY ———————→ HARD

TEACHER PREP
CONCEPT LEVEL
CLEAN UP

MATERIALS

- water
- 500 mL beaker
- 4 envelopes of unflavored gelatin
- 2 L bowl
- stirring rod or spoon
- red food coloring
- small container with a lid
- 3 L bowl
- 4 clay bricks
- pegboard, 40 × 60 cm, with 5 mm holes
- large tray or cookie sheet
- syringe (found in pet stores)
- rubber tubing
- large knife

Advance Preparation

- Prepare the gelatin for the volcano model by mixing 500 mL of cool water with the contents of four envelopes of unflavored gelatin in the 2 L bowl. Stir for 30 seconds. Boil 1.5 L of water in a saucepan on a hotplate. Add 1.5 L of boiling water,

and stir until the gelatin is dissolved. Refrigerate for at least 3 hours or overnight.

- Prepare the "magma" in the small container by mixing just enough water with red food coloring to make a very dark liquid. Set aside.

What to Do

1. Fill the 3 L bowl about halfway with hot water. Remove the gelatin from the refrigerator, and loosen it from the bowl by dipping the bowl briefly in the larger bowl of hot water.

2. Arrange the four bricks in a rectangle on the table so that they will be able to support the pegboard. Lay the pegboard on top of the bricks. Slide a tray or cookie sheet under the setup for collecting any drips, as shown in the diagram.

3. Unmold the gelatin by tipping the bowl over onto the center of the pegboard and lifting the bowl. Do this very carefully so that the gelatin cast won't develop cracks; a few small cracks are acceptable. The gelatin cast will spread and settle. It should resemble a colorless to milky volcano.

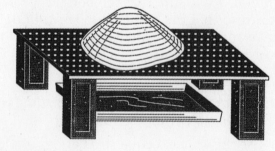

continued...

Vicky Farland
Crane Jr. High
Yuma, Arizona

4. Fill the syringe with your prepared "magma." Remove any air bubbles from the syringe by holding the syringe upright and squirting out a small amount of the liquid. Air tends to fracture the gelatin. Attach tubing to the syringe.

5. Tell students that you are now going to simulate the movement of magma through a volcano.

6. Insert the other end of the tubing into the center of the gelatin cast through a hole in the pegboard. Inject the red water very slowly, at a rate of about 20 cc/min, and have students watch carefully.

7. Ask students to describe what they observed. *(If the gelatin is fractured, the red dye will flow out of the fractures. If there are no fractures, the dye should flow out equally in all directions.)*

8. Refill the syringe several times. Compare the movement of the magma each time.

9. Use the knife to slice open the volcano, and view the cross-section. Tell students that in the gelatin, just as in a volcano, the magma will follow the path of least resistance. When there is a fracture in the rock on the Earth's surface, the magma will flow out of the fractures. Otherwise, the magma flows in all directions.

Explanation

Lava is magma, or molten rock, that seeps through the Earth's surface. Underground, it travels through cuts or fractures in the rock. Vertical fractures in the surrounding rock fill with the magma, creating large sheets of moving molten rock called dikes. Similarly, in this demonstration the cracks in the gelatin fill with the red liquid and open as the pressure from the liquid increases.

Going Further

Prepare gelatin in a bread pan to compare the flow of dye in the bread pan with that in the round bowl. In this case, the magma will flow horizontally rather than vertically.

EARTH SCIENCE

MAKING MODELS

When It Rains, It Fizzes

Purpose
Students are introduced to the concepts of acid precipitation and chemical weathering.

Time Required
5–10 minutes

Lab Ratings

EASY ———————→ HARD

TEACHER PREP
CONCEPT LEVEL
CLEAN UP

MATERIALS

- calcium carbonate tablet or piece of chalk
- 250 mL beaker
- 200 mL of vinegar
- protective gloves

Safety Information
Vinegar is an acid; it should be handled with caution. Wear safety goggles and protective gloves at all times. Clean up spills immediately.

What to Do
1. Place the calcium carbonate tablet in the beaker. Tell students that the tablet represents limestone, a rock common to the Earth's crust and one that is used in many statues and buildings.

2. Fill the beaker with 200 mL of vinegar. Tell students that vinegar is a weak acid; it represents acid rain in this demonstration.

3. Students will notice fizzing and foaming as a chemical reaction takes place

in the beaker. Ask students to describe what is happening. *(The vinegar is causing the calcium carbonate pill to separate and dissolve.)*

Explanation
Industries, cars, and machinery that consume fossil fuels are largely responsible for the pollutants that cause acid precipitation. Acid precipitation forms when gases in the air, primarily sulfur and nitrogen dioxides, react with water droplets. The products of these reactions, nitric and sulfuric acids, hasten the decomposition of rocks and minerals during a process called *chemical weathering*. The weathering of limestone, marble, and other rocks by acid precipitation is particularly noticeable in areas where temperatures are high and rainfall is abundant.

All rain is mildly acidic due to a reaction between carbon dioxide and water in the atmosphere. Therefore, even in the absence of pollution, any precipitation induces mild chemical weathering.

Discussion
Use the following questions as a guide to encourage class discussion:

- How did this demonstration simulate real-world processes? *(The demonstration simulates the weathering effects acid precipitation can have on rocks, statues, and buildings.)*

- **Critical Thinking** How do you think the reaction could have been slowed down? *(Sample answer: The reaction would have slowed down if concentration of the acid had been weaker or if the vinegar had been colder.)*

CLASSROOM TESTED & APPROVED

Jerry Taylor
Santa Maria Middle School
Phoenix, Arizona

Between a Rock and a Hard Place

Purpose

Students see some of the factors involved in erosion and learn how some unusual land formations may have formed.

Time Required

15–20 minutes

Lab Ratings

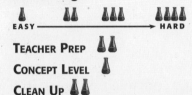

EASY ————→ HARD

TEACHER PREP 🧪🧪
CONCEPT LEVEL 🧪
CLEAN UP 🧪🧪

MATERIALS

- small amount of pottery clay (not modeling clay or another oily clay)
- large aluminum pie plate
- 5 or 6 pebbles or pennies
- hose with sprayer attachment

Advance Preparation

This demonstration should be conducted outdoors. Find a suitable location for spraying water from a hose.

You may wish to show students photographs of landforms that are created by the method demonstrated here. Such formations are located in Bryce Canyon National Park and Canyonlands National Park, both in Utah.

What to Do

1. Form a flat brick about 15 × 5 × 3 cm out of the clay.

2. Place the clay in the aluminum pie plate.

3. Firmly press the pebbles into the top of the clay brick. Space the pebbles evenly along the top of the clay brick.

4. Ask students to predict what will happen if you direct a stream of water from the hose at the clay and pebbles. *(Expected answer: The pressure of the water will wash the pebbles away.)*

5. Now spray the brick with water from the hose. Begin by using a gentle spray, and increase the pressure until the brick begins to wash away.

Discussion

Use the following questions as a guide to encourage class discussion:

- What happens to the clay when the hose is turned on? *(The water washes away some of the clay.)*

- What do you observe happening to the clay underneath the pebbles? *(The clay does not wash away as quickly.)*

- Why do you think this happened? *(The pebbles protected the clay underneath them from the water.)*

- What would the Earth look like where this erosion occurred for thousands of years? *(It would look similar to the formation common in Bryce Canyon National Park.)*

EARTH SCIENCE

Jane Lemons
Western Rockingham
Middle School
Madison, North Carolina

Rising Mountains

Purpose

Students discover that mountains rise as they erode.

Time Required

15–20 minutes

Lab Ratings

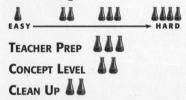

TEACHER PREP

CONCEPT LEVEL

CLEAN UP

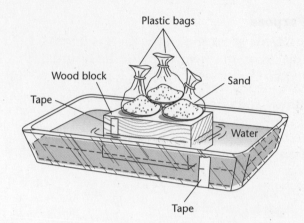

MATERIALS

- transparent, shallow plastic container
- tap water
- wooden block—about 5 × 10 × 10 cm
- 3–4 plastic sandwich bags
- sand (enough to fill sandwich bags)
- masking tape
- set of colored waterproof markers

What to Do

1. Fill the plastic container about two-thirds full with water. Place the block in the center of the water.

2. Fill each plastic bag with sand, and then place the bags on the block.

3. Tell students that the block and the bags represent a mountain. The mountain is a formation on the Earth's outermost layer, the crust. The Earth's crust is constantly moving in relation to the mantle, which is represented by the water.

4. Use masking tape and a marker to indicate the water level on the wooden block

and the plastic container, as shown in the diagram. Tell students that these lines will measure the original position of the mountain on the Earth's surface.

5. Tell students that erosion is constantly removing rocks and sand from the sides and top of mountains. Rocks and soil roll, slide, blow away, and wash away because they are being eroded by wind, rain, rivers, and gravity. Ask students to imagine that the sandbags are small rocks and soil being eroded from the mountainside. Ask: What would you expect to happen if the sandbags were removed (or if the soil were eroded away)? *(The wooden block would rise and float higher.)*

6. Gently remove a bag from the wooden block. With another color of marker, indicate the water level again. Compare this mark with the original water level.

7. Repeat the process several times to illustrate that the motion is slow and gradual. Mountains rise slowly and gradually over thousands of years.

Peggy Belt
West Hardin Middle School
Cecillia, Kentucky

Fowl Play

Purpose

Students learn how an oil spill and its cleanup can affect birds and other wildlife.

Time Required

15–20 minutes

Lab Ratings

EASY ——————→ HARD

TEACHER PREP
CONCEPT LEVEL
CLEAN UP

MATERIALS

- 3 small, transparent bowls
- tap water
- 50 mL beakers (2)
- 20 mL of vegetable oil
- 40 mL of powdered detergent
- 3 feathers (5–7 cm long)
- spoon

What to Do

1. Fill the bowls three-fourths full of water. The first bowl should contain only tap water. Add 20 mL of oil to the second bowl. Stir 20 mL of detergent into the third bowl.

2. Gently lay a feather on the surface of the water in the first bowl. Remove the feather, and blow on it. Put it back in the water. Ask students to describe what happened. *(Expected answer: The feather is dry, smooth, and clean. It floated in the water.)*

3. Gently lay the second feather on the oil in the second bowl. Remove the feather and blow on it. Put it back on the oil. Ask students to describe what happened. *(Expected answer: It is matted and sticky. It floated when it was put back on the oil.)*

4. Add 20 mL of detergent to the second bowl, and stir. Ask students to describe

what happened to the oil. *(Expected answer: The detergent caused the oil to break up into tiny bubbles, and some of the oil sank.)*

5. Gently place the third feather in the third bowl, moving it around so that it is coated with detergent. Ask the students to describe what happens. *(Expected answer: The feather became soaked with water, and it sank.)*

Discussion

Use the following questions as a guide to encourage class discussion:

- Based on the demonstration, how would a bird's feathers perform in uncontaminated water? *(Expected answer: Normally, uncontaminated feathers stay dry and clean, allowing birds to float in water and fly through the air.)*

- How would an oil spill affect a bird's ability to fly? *(Expected answer: The oil would cause a bird's feathers to become sticky and to clump together. The bird would not be able to fly efficiently if at all.)*

- Is detergent an effective method for cleaning up oil spills? *(The detergent broke up the oil. Unfortunately, it also made the feather sink. Therefore, detergent is an effective method for breaking up oil, but it may be harmful to the birds that come in contact with it.)*

Going Further

Suggest that students research new methods for cleaning up oil spills that do not harm wildlife.

Daniel Bugenhagen
Yutan Jr.–Sr. High
Yutan, Nebraska

► EARTH SCIENCE

MAKING MODELS

Spin Cycle

Purpose

Students observe the Coriolis effect and learn how the Earth's rotation affects ocean currents.

Time Required

10–15 minutes

Lab Ratings

EASY ➙ HARD

TEACHER PREP

CONCEPT LEVEL

CLEAN UP

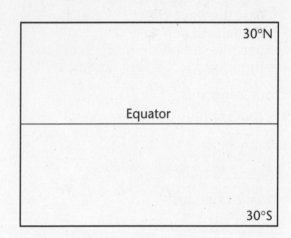

MATERIALS
• sheet of white paper • pencil • metric ruler • transparent tape • coffee can • duct tape • felt-tip pen

What to Do

1. Tell students that ocean currents exist because warm air and warm water at the equator expand and move toward the poles due to the process of convection.

2. Tell students that this demonstration will show how the Earth's rotation affects ocean currents.

3. Begin by folding the paper in half lengthwise. Open the paper, and use the pencil and ruler to draw a straight line along the crease.

4. Turn the paper so that the line runs horizontally. Label this line "Equator." Write "30°N" in the top right corner and "30°S" at the bottom right corner of the paper, as shown in the diagram.

5. Fold the paper again, and tape it around the can so that 30°N is at the top.

6. Place the can near the edge of a table. Use the duct tape to secure the ruler vertically against the edge of the table so that the top of the ruler is even with the top of the can. See the diagram below.

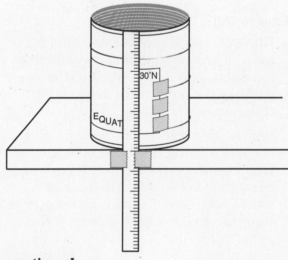

continued...

Michael E. Kral
West Hardin Middle School
Cecillia, Kentucky

7. Rotate the can so that about 3 cm of the right edge of the paper extends past the right edge of the ruler. Hold the felt-tip pen against the top of the paper and next to the right edge of the ruler.

8. Call on a student to turn the can quickly in a counterclockwise direction about one quarter turn while you draw a line on the paper straight down the edge of the ruler. Draw an arrowhead at the end of this line to indicate the direction in which you drew the line.

9. Remove the paper from the can, turn the paper over, and tape the paper to the can so that 30°S is showing at the bottom. Return the can to its original position on the table, and rotate the can so that about 3 cm of the right edge of the paper extends past the right edge of the ruler.

10. Hold the pen against the bottom of the paper and next to the right edge of the ruler. Again ask a volunteer to rotate the can counterclockwise as you draw a line on the paper straight up the edge of the ruler.

11. Draw an arrowhead at the end of this line to indicate the direction in which you drew the line. Remove the paper from the can and unfold the paper.

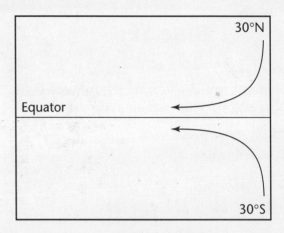

12. Observe the direction of the two arrows. Notice that both arrows curve to the left and toward the equator.

Explanation

The ocean's surface currents, driven by the prevailing winds, would flow in a north-south loop were it not for the rotation of the Earth. The Earth's rotation causes the currents to bend, or deflect. This deflection is called the Coriolis effect.

In this activity, turning the can represented the rotation of the Earth. The lines you drew along the ruler would have been straight up and down if the can had not been turned. Similarly, ocean currents and winds would follow a straight path if the Earth were not rotating on its axis.

EARTH SCIENCE

Blue Sky

Purpose

Students investigate why the midday sky is blue and why sunsets are red.

Time Required

10–15 minutes

Lab Ratings

EASY ———————————→ HARD

TEACHER PREP
CONCEPT LEVEL
CLEAN UP

MATERIALS

- empty aquarium
- tap water
- flashlight or slide projector
- powdered milk or powdered coffee creamer
- blank white card

Advance Preparation

Prepare a solution as follows to simulate the atmosphere.

- Fill the aquarium with water, and shine a beam of light into the water so that the beam is parallel to the length of the aquarium.

- Stir a small amount of powdered milk or powdered coffee creamer into the water. As you do so, the beam of light will become more noticeable.

- Continue to add powder until the beam is clearly visible from across the room. Do not add too much powder! Remove the beam of light.

Before performing this demonstration for the class, you may wish to turn out the lights and allow time for everyone's eyes to adjust to the darkness.

HELPFUL HINT

If students have difficulty seeing the colors, try dimming the lights in the classroom further or using a narrower beam of light. One way to produce a narrow beam is to use a hole punch to create a hole in an unexposed black slide or in an index card cut to the size of a slide. Then place the slide or index card in a slide projector and focus the projector to obtain a sharp, narrow beam.

What to Do

1. Ask students to offer an explanation of why the sky is often blue at midday and orange-red at sunset. *(Accept all reasonable responses.)* Tell them that you will now perform a demonstration to help them determine the answer.

2. Shine the beam of light through the tank from one end of the aquarium. Encourage a few students at a time to walk around the tank and observe the beam from different positions.

3. Ask a volunteer to hold up a white card at the other end of the tank in the path of the beam. This will allow students to observe the color of the light as it leaves the water.

4. After students have made their initial observations, add more powder to the water. Students should observe that the colors along the beam change from blue-white to yellow-orange.

5. Offer students a chance to revise their previous explanations based on what they have just observed.

continued...

Michael E. Kral
West Hardin Middle School
Cecillia, Kentucky

Explanation

White light is scattered by dust in the atmosphere as it travels from the sun to Earth. Light that has shorter wavelengths (blue) is more strongly scattered by the atmosphere than light that has longer wavelengths (red). In fact, blue light is scattered about 10 times more than red light!

This means that blue light is scattered out of a beam of white light faster than red light is scattered. The blue light is scattered into the atmosphere, giving the sky its usual color.

The color of a beam of white light that travels a relatively short distance in the atmosphere, such as from a point directly overhead, does not change very much.

However, at sunset, light travels a long distance through the atmosphere. When the light finally reaches our eyes, more of the blue light has been scattered compared to the red light, so the beam appears red. The more dust particles there are in the atmosphere, the stronger the scattering is and the redder the sunset is.

Discussion

Use the following questions as a guide to encourage class discussion:

- What does the light beam represent in this model? *(The light beam represents the sun's rays.)*

- What does the milky water represent? *(The milky water represents the atmosphere.)*

- Why did I add powder to the water? *(Powder was added to the water to simulate the molecules and dust in the atmosphere that scatter light.)*

- Where in the sky is the sun when you are most likely to see an orange-red sky? *(The sun is low on the horizon in the evening.)*

- **Critical Thinking** Based on this model, do you think that the sun's light travels through more or less of the atmosphere when it is low on the horizon? Why? *(I think that the sun's light has to travel through more of the atmosphere when it is low on the horizon; in this demonstration, white light traveled through more of the milky water to produce the orange-red light.)*

Going Further

If students are having trouble understanding the concept, use a compact disc to demonstrate how another material can separate the wavelengths of light. When the disc is held and rotated, colors appear on the disc's surface. This happens because the shape of the surface causes the light that strikes the disc to separate into colors. Put the CD into a CD case, and notice the change in the CD's ability to separate wavelengths of light. The CD case acts like atmospheric dust particles, scattering light before it gets to the disc.

EARTH SCIENCE

When Air Bags Collide

Purpose

Students observe the motion of two colliding bags to reinforce their understanding of the behavior of air masses at warm and cold fronts.

Time Required

5–10 minutes

Lab Ratings

EASY ——————————→ HARD

TEACHER PREP

CONCEPT LEVEL

CLEAN UP

MATERIALS

- 2 clean, different-colored pillowcases
- marker
- foam packing peanuts or other low-density material, enough to fill 2 pillowcases
- large bag of dried beans
- 2 large rubber bands

Advance Preparation

Construct two large, flat bags to represent a warm air mass and a cold air mass. Use a marker to label one pillowcase "warm" and one pillowcase "cold." Fill both pillowcases with the low-density material. Then add a denser material, such as dried beans, to the pillowcase labeled "cold." Enough beans should be added so that the cold air mass does not move easily when pushed by the warm air mass. Use the rubber bands to close the open ends of each pillowcase to form a bag.

What to Do

1. Lay both bags side by side on a table in front of the class. Point out the bag that represents the warm air mass and the bag that represents the cold air mass. Remind students that cold air masses are denser than warm air masses.

2. Explain that a front is the boundary between two air masses of different temperatures. Call on students to show where the front exists in your model.

3. Ask students, "What happens at a cold front when a cold air mass arrives at a warm air mass?" *(The dense, cold air compresses and lifts the warm, less dense air.)* Demonstrate what happens at a cold front by sliding the cold air mass against the warm air mass.

4. Ask students, "What happens at a warm front when a warm mass collides with a cold air mass?" *(The dense, cold air is not moved by the warm air. Instead, the warm air slides or rolls over the cold air.)* Demonstrate what happens at a warm front by sliding the warm air mass against the cold air mass.

Explanation

This demonstration is a good illustration of the effects of density and inertia on the motion of air masses. Because cold air is denser than warm air, it easily displaces warm air but is not easily displaced by warm air.

The change in weather we experience when passing through a cold front is much more abrupt than the change we experience when passing through a warm front. At a warm front, the warm air slides over the cold air, so the transition is very gradual. At a cold front, the cold air compresses the warm air, so the transition is more abrupt.

Peggy Belt
West Hardin Middle School
Cecillia, Kentucky

It's Raining Again

Purpose

Students observe the water cycle in action.

Time Required

10–15 minutes

Lab Ratings

EASY ———————→ HARD

TEACHER PREP

CONCEPT LEVEL

CLEAN UP

MATERIALS

- 2 L clear plastic soda bottle with cap
- scissors
- 1 L beaker
- hotplate
- tap water
- 7–10 pieces of ice

What to Do

1. Cut off approximately one-fourth of the top of a soda bottle. Save the top.

2. Boil 500 mL of water in a beaker on a hotplate. Pour the boiling water into the base of the bottle.

3. Insert the top section of the bottle upside down into the base of the bottle, as shown in the diagram below.

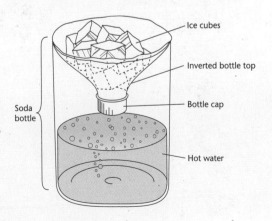

- Ice cubes
- Inverted bottle top
- Bottle cap
- Soda bottle
- Hot water

4. Add the ice to the bowl formed by the top of the bottle.

5. Tell students to sit back and watch what happens. *(Students should observe clouds forming and water droplets falling back into the hot water.)*

Explanation

As water vapor escapes from the boiling water and mixes with the cold air around the ice, clouds form and drops of liquid water precipitate.

This is similar to the formation of clouds and rain in the atmosphere. Water evaporates from oceans, rivers, and lakes. As pockets of warm air, called *thermals*, rise from the Earth's surface, they carry the water vapor up. As the thermals get farther from the Earth's surface, the air within them cools and water vapor condenses to form clouds. When enough water accumulates in these clouds, rain falls.

continued...

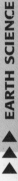

EARTH SCIENCE

Daniel Bugenhagen
Yutan Jr.–Sr. High
Yutan, Nebraska

Discussion

Use the following questions as a guide to encourage class discussion:

- How does this demonstration simulate the water cycle? You may wish to have a volunteer draw the cycle on the board. *(Steam rises from the water in the bottle until it encounters the cool surface above. There, it condenses into droplets, which fall back into the bottom of the bottle. The steam in this lab represents water vapor rising off the surface of the Earth. The vapor condenses into clouds when it encounters cool temperatures at high altitudes. Clouds accumulate water droplets until the drops are heavy enough to fall as rain.)*

- What is the function of the ice in this model? *(The ice cooled the air in the top of the bottle, promoting condensation.)*

- What is the function of the inverted bottle top? *(The inverted bottle top gave the water a surface on which to condense.)*

MAKING MODELS

How Humid Is It?

Purpose
Students learn about relative humidity.

Time Required
10–15 minutes

Lab Ratings
EASY ——————————→ HARD

TEACHER PREP
CONCEPT LEVEL
CLEAN UP

MATERIALS

- 2 Celsius thermometers
- piece of gauze
- rubber band
- tap water
- transparent tape
- piece of cardboard

What to Do

1. Tell the class that you will make an object called a *psychrometer* to measure relative humidity. Explain that relative humidity tells how much moisture is in the air. Warm air can hold more moisture than cold air. Air is said to be saturated when it cannot hold any more moisture at a given temperature. Relative humidity is a measurement of the percentage of saturation at a given temperature.

2. Secure the gauze to the bulb of one of the thermometers with a rubber band. Wet the gauze by dipping it in water. Tell students that this is a wet-bulb thermometer and that the other is a dry-bulb thermometer.

3. Securely tape the two thermometers over the edge of a desk or table. Be sure that they are secure and that no one will bump them.

4. Vigorously fan the thermometers with the cardboard. The temperature of the wet-bulb thermometer will drop as evaporation cools the wet gauze. When the temperatures of both thermometers stabilize, stop fanning.

5. Read the temperature of each thermometer, and record your findings on the board in the form of a formula: Dry temperature–Wet temperature = x.

6. Have students subtract the wet bulb's reading from the dry bulb's, and record the difference, x.

7. Use the Relative Humidity Table on the next page to determine the humidity of the classroom.

Discussion
Use the following questions as a guide to encourage class discussion:

- How did evaporation affect the wet bulb's temperature reading? *(As water evaporated from the bulb's surface, the temperature reading dropped.)*

- How might relative humidity influence evaporation? *(The difference in temperature between the wet and dry bulb readings is smaller when the relative humidity is high. Therefore, as the relative humidity increases, less moisture is able to evaporate.)*

- **Critical Thinking** What would the relative humidity be if the wet and dry bulbs read the same temperature? *(They would read the same if the air were saturated and unable to absorb any more moisture. The relative humidity would then be 100 percent.)*

continued...

Janet Nilsson
Rosemont Middle School
Ft. Worth, Texas

EARTH SCIENCE

Relative Humidity Table

Dry bulb °C	Relative humidity (percent)									
10	88	77	66	55	44	34	24	15	06	0
11	89	78	57	56	46	36	27	18	9	0
12	89	78	68	58	48	39	29	21	12	0
13	89	79	69	59	50	41	32	22	15	7
14	90	79	70	60	51	42	34	26	18	10
15	90	80	71	61	53	44	36	27	20	13
16	90	81	71	63	54	46	38	30	23	15
17	90	81	72	64	55	47	40	32	25	18
18	91	82	73	65	57	49	41	34	27	20
19	91	82	74	65	58	50	43	36	29	22
20	91	83	74	67	59	53	46	39	32	26
21	91	83	75	67	60	53	46	39	32	26
22	92	83	76	68	61	54	47	40	34	28
23	92	84	76	69	62	55	48	42	36	30
24	92	84	77	69	62	56	49	43	37	31
25	92	84	77	70	63	57	50	44	39	33
x:	1	2	3	4	5	6	7	8	9	10

DISCOVERY LAB

Refraction Action

Purpose

Students learn why the sun can be seen before it rises above the horizon.

Time Required

10–15 minutes

Lab Ratings

EASY ————————→ HARD

TEACHER PREP

CONCEPT LEVEL

CLEAN UP

MATERIALS

- clean, clear 2 L soda bottle with a lid
- tap water
- metric ruler
- 3–4 books
- small rubber ball or a piece of clay the size of a walnut

HELPFUL HINT

You may want to bring in enough materials to set up several viewing stations.

What to Do

1. Fill the bottle with water until it overflows, and tightly screw on the lid.

2. Lay the bottle on its side about 30 cm from the edge of a table. Stack books between the edge of the table and the bottle so that about one-fourth of the bottle is visible above the books. Lay the rubber or clay ball on the table about 10 cm behind the bottle.

3. Have pairs of students take turns kneeling down in front of the books and looking straight through the bottle of water. If the clay ball is not visible, move it to a new position.

4. While each pair of students is looking through the bottle at the same level, remove the bottle from the table. Students should not be able to see the clay ball.

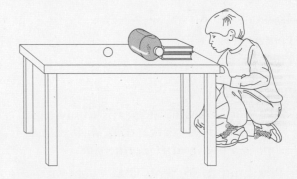

Explanation

Students could see the clay ball through the bottle because the light was bent, or *refracted*, as it passed through the water. Refraction occurs when light waves cross the boundary between two substances of different densities. The Earth's atmosphere refracts light in the same way the bottle of water does. As light travels from the sun to the Earth and passes through the boundary between outer space and the atmosphere, the light waves bend. This causes the sun to be visible before it actually rises in the morning and for a few minutes after it sets in the evening.

Janet Nilsson
Rosemont Middle School
Ft. Worth, Texas

EARTH SCIENCE

Can You Vote on Venus?

Purpose

Students calculate their age in years on other planets and develop a better understanding of planetary motion around the sun.

Time Required

10–15 minutes

Lab Ratings

EASY ──────────→ HARD

TEACHER PREP

CONCEPT LEVEL

CLEAN UP

What to Do

1. Ask students what it means to say that someone is 10 years old. *(It means that a person has lived 10 Earth years.)*

2. Point out to students that a year on Earth is the time it takes for the Earth to complete one revolution around the sun. Other planets revolve around the sun at different speeds, so a year is not the same on every planet.

3. Tell students they will calculate their age in years on the other planets. First have them multiply their age on Earth by 365 to find their age in Earth days. Then have them divide their age in Earth days by the number of Earth days that make up a year on a particular planet. The following is an example for a 25-year-old person on Mercury:

$$25 \text{ Earth years} \times \frac{365 \text{ Earth days}}{1 \text{ Earth year}}$$

$$= 9{,}125 \text{ Earth days}$$

Mercury completes one revolution around the sun in 88 Earth days.

$$9{,}125 \text{ Earth days} \div \frac{88 \text{ Earth days}}{1 \text{ Mercury year}}$$

$$= 104 \text{ Mercury years}$$

A person who is 25 years old on Earth is 104 on Mercury!

The number of Earth days or Earth years for each planet is given in the chart below. Have students record their answers in their ScienceLogs.

Table of Planetary Years

Planet	Length of year
Mercury	88 Earth days (0.24 Earth years)
Venus	225 Earth days (0.62 Earth years)
Earth	365 Earth days (1.00 Earth years)
Mars	687 Earth days (1.88 Earth years)
Jupiter	4,383 Earth days (12 Earth years)
Saturn	10,775 Earth days (29.5 Earth years)
Uranus	30,681 Earth days (84 Earth years)
Neptune	60,266 Earth days (165 Earth years)
Pluto	90,582 Earth days (248 Earth years)

Discussion

After completing the activity, ask students to look for trends in their responses. Point out that they would be older on Mercury and Venus and younger on all of the other planets. This is because Mercury and Venus revolve around the sun faster than the Earth does, while Earth revolves around the sun faster than all of the remaining planets do. Students should be led to discover that the farther a planet is from the sun, the longer it takes the planet to revolve around the sun.

continued...

Use the following questions as a guide to encourage class discussion:

- Would you actually be older on Mercury? *(No. The same amount of time would pass, no matter what planet you were on. During time spent on Mercury, you simply would have completed more trips around the sun.)*

- Could you vote on Venus? *(Assuming that the voting age is 18, any student over 4,052 Earth days old—11.1 Earth years— would be of voting age on Venus.)*

Explanation

Time is the same on every planet. If we were to create a calendar for all of the planets in the solar system, they would all be different because each planet is a different distance from the sun. One years is always a full trip around the sun. Planets that are close to the sun can make their revolutions quickly, so a year on those planets is relatively short. Planets farther from the sun take longer to make a full revolution and thus have relatively long years.

EARTH SCIENCE

MAKING MODELS

Crater Creator

Purpose

Students learn about impact craters and explore the factors that contribute to a crater's appearance.

Time Required

15–20 minutes

Lab Ratings

EASY ————————→ HARD

TEACHER PREP

CONCEPT LEVEL

CLEAN UP

MATERIALS

- several sheets of newspaper
- cardboard box or plastic pan at least 10 cm deep
- all-purpose flour
- piece of cardboard, about 10 × 20 cm
- sifter or sieve
- tempera paint powder or chocolate-drink-mix powder
- small, medium, and large marbles or ball bearings
- metric measuring tape

USEFUL TERMS

ejecta blanket
a layer of ejecta (or debris) that lies like a blanket close to the impact crater

raised rim
a ring of debris deposited around the crater's edge

rays
material ejected away from the craters, seen as long lines

What to Do

1. Spread the newspapers on the floor. This can get messy. Place the box or pan on the floor in the center of the newspapers, and add enough flour to fill the container so that the flour is 6–8 cm deep. Smooth the surface with a piece of cardboard. Use the sieve or sifter to dust the surface of the flour with a light coating of tempera paint powder or chocolate-drink-mix powder.

2. Tell students that you will be simulating impact craters on lunar or planetary surfaces. The marbles will represent meteors or asteroids, and the flour simulates the surface.

3. After each impact, ask students to observe the flour's surface for the features of impact craters: *ejecta, raised rim,* and *rays.* Ejecta surrounds a crater; it is a layer of debris that spreads on impact. The raised rim is the edge of the crater itself. This is the circle of debris that marks the perimeter of the impact. Rays are formed as material is ejected from the surface and radiates away from the impact.

4. Drop a small marble or ball bearing onto the flour from a height of 50 cm, 100 cm, and 150 cm. Then throw the marble from a height of 100 cm. Finally, throw the marble into the flour from a different angle.

continued...

Daniel Bugenhagen
Yutan Jr.–Sr. High
Yutan, Nebraska

5. After several crater impacts, you may need to refresh the surface by mixing the powder into the flour. Then create another smooth surface with the cardboard, and finally sift more powder over the mixture.

6. Repeat step 4 with the medium and large marbles.

7. Ask students to compare the results of the different-sized "asteroids." *(The size of the projectile as well as its velocity and angle of impact affect the appearance of the crater. The bigger marbles made deeper craters. Throwing the marble harder also made the crater deeper.)*

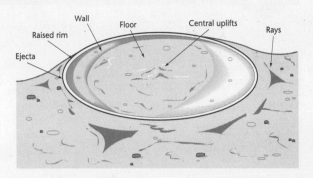

Discussion

Use the following questions as a guide to encourage class discussion:

- Which angle made the biggest ejecta blanket? *(Accept all reasonable responses.)*

- Which objects made the largest craters? *(Accept all reasonable answers.)*

- **Critical Thinking** The Barringer Crater, in Arizona, was formed when a meteor 30–50 m in diameter struck the Earth 50,000 years ago. It made a crater 1,200 m across and 200 m deep. How did a meteor that size make a crater so much

larger than itself? *(The size of a crater depends not only on the size of the meteor but also on its velocity. The Barringer meteor was traveling at a very high velocity and hit the Earth with tremendous force, forming a crater many times larger than itself.)*

Explanation

Craters typically form where lava and ash are ejected out of a volcano or when a projectile strikes a surface of a moon or planet. This demonstration focuses on the latter kind, which are called impact craters. Impact craters form when a projectile, such as an asteroid, a meteoroid, or a comet, strikes the surface of a planet or moon. The size of the projectile as well as its velocity and angle of impact affect the appearance of the crater.

EARTH SCIENCE

Space Snowballs

Purpose
Students learn about the composition of comets in this exciting demonstration.

Time Required
35–45 minutes

Lab Ratings
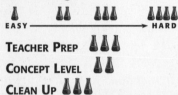

EASY ⟶ HARD

TEACHER PREP ♦♦♦
CONCEPT LEVEL ♦♦
CLEAN UP ♦♦♦

MATERIALS

- pair of waterproof, cold-resistant gloves
- apron
- 2.3 kg of dry ice
- several sheets of newspaper
- 2 trash bags
- large bowl or small pot
- tap water
- box of corn starch
- drops of ammonia (or a few sprays of window cleaner)
- handful of fine-grained dirt
- mixing spoon
- cloth towel

Safety Alert
Dry ice can cause frostbite, so be sure to wear cold-resistant gloves and an apron when handling it. When breaking the dry ice into pieces, be sure to wear goggles to prevent small pieces from harming your eyes. Should chemicals contact your eyes, immediately flush with plenty of water. The comet and any remaining dry ice can be safely allowed to sublimate in a sink or container in a well-ventilated room, provided that students are not allowed access.

Advance Preparation
Frozen CO_2, or dry ice, is available from many commercial ice vendors. Just look in the phone book under *ice*. Many ice cream shops receive their goods packed in dry ice, so they might also be a convenient (and possibly free) source of dry ice. You will need about 2.3 kg of dry ice for each nucleus. Purchase the ice the night before the activity and store it in a refrigerator or a cooler. Line the base of your storage container with about 3 cm of newspaper to prevent the container from cracking.

What to Do

1. Begin the demonstration by lining the bowl with a trash bag and pouring about 1 pint of water into the bowl. Add the corn starch, ammonia, and some of the dirt. Mix slightly.

2. Place the other trash bag on the floor, and put on the gloves. Wrap the dry ice in a cloth towel, and place it on the trash bag on the floor. Use a hammer to reduce the dry ice to a powder. Then gradually pour the dry ice powder into the water, mixing as you pour.

3. Tell students to watch carefully. Soon they should see a vapor. Stir the mixture until the dry ice, water, and other ingredients form a thick slush. Keep stirring for a few seconds as the mixture thickens.

continued...

Richard Petitt
Bardstown Middle School
Bardstown, Kentucky

4. Use the trash bag to lift the slush away from the sides of the bowl. Pack the slush into a ball. Keep packing until it solidifies into a ball. This ball is a model of a comet nucleus.

5. Peel the trash bag from the lump. Scatter some more dirt over the ball. While turning the ball, pour some of the remaining water over it so that a layer of ice forms over the entire ball.

6. Have students observe the behavior of the comet-nucleus model. It will hiss and pop as frozen carbon dioxide sublimates (goes from a solid state directly into a gas) and forces its way through weak spots in the water-ice crust. This creates small jetting forces that would cause a real comet nucleus to spin, to alter its orbit slightly, or to split apart.

Explanation

Comets are sometimes called dirty snowballs because they are a collection of stone or metallic debris held together by frozen water and gases. For hundreds of years, the arrival of a comet in the sky was thought to be an omen of famine and war. Although these superstitions are no longer common, comets are still watched closely by scientists and amateur astronomers alike. Many questions about the origin and movement of comets remain, but their makeup and structure are now well known.

The materials in this lab are either actual components of a comet or convenient analogous ones. The dry ice is frozen carbon dioxide. Water, ammonia, organic (carbon-based) molecules, and silicates are all present on comet nuclei. In the activity, the dirt provided organic molecules, and sand is a silicate.

Discussion

Students may be interested to know the following fun facts about comets:

- During most of its orbit, a comet is far from the sun. During this time, it is merely a nucleus. Only when it approaches the sun does a comet have a head and a tail.

- The nucleus of a typical comet is usually no more than a few kilometers across, although some comets have nuclei as large as 161 km across.

- When a comet nears the sun, the ice in it *sublimates*, or changes directly from a solid to a gas. The big vapor cloud produced forms the head of the comet.

- Dusty silicate particles and gases stream from the comet's head to form the tail. This tail is always directed away from the sun because there is a constant stream of particles flowing from the sun as *solar wind*. This solar wind pushes the loose silicates on the comet away from the sun.

- The tail of a comet can stream for as many as 161 million kilometers!

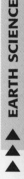

EARTH SCIENCE

DISCOVERY LAB

Where Do the Stars Go?

Purpose

Students observe that stars shine in the sky even in the daytime, when it is too bright to see them.

Time Required

About 5–10 minutes

Lab Ratings

EASY ————————————→ HARD

TEACHER PREP
CONCEPT LEVEL
CLEAN UP

MATERIALS

- index card
- paper-hole punch
- envelope, slightly larger than the index card
- 2 flashlights
- overhead projector

What to Do

1. Punch 7–10 holes in the index card, and insert the card into the envelope.

2. Hold up the envelope, and shine a flashlight or overhead projector at the card from behind. Tell students to imagine they are seeing stars in the night sky.

3. Ask for a volunteer to hold the other flashlight, and shine the light at the front of the card. The stars will disappear.

4. Have your assistant turn off his or her flashlight. The stars will reappear.

Discussion

Use the following questions as a guide to encourage class discussion:

- What did the second flashlight represent? *(It represented the sun.)*

- What happened to the stars when the sun appeared? *(They were no longer visible.)* Why? *(The stars were still present in the sky, but the bright light from the sun prevented them from being observed.)*

Explanation

The stars are always shining. When the sun is below the horizon, most of the light reaching our eyes comes from the stars, making them visible. This corresponds to step 2 at left.

When the sun is above the horizon, however, light from the sun scatters in the Earth's atmosphere, causing the sky to appear blue. It is this scattered light that obscures the stars, which are too dim to outshine the bright daytime background. Instead, the light from the stars blends with the bright light from the sun. It is difficult to distinguish a star in the daytime without very powerful instruments. This corresponds to step 3 above. Point out that the light from the flashlight reflects off the white card. Thus, the card plays the role of the Earth's atmosphere.

To emphasize this point, ask students, "Could you see the stars during the daytime if you were on the moon, where there is no atmosphere?" The answer is that stars are visible during the daytime if you were on the moon. This can be demonstrated by taking the second flashlight and shining it at the observer, just as the sun would shine on an observer on the moon. The observer will most likely have to shield his or her eyes from the direct light of the sun, but by doing so the stars should become visible.

CLASSROOM TESTED & APPROVED

Patricia Marks
San Marcos Middle School
San Marcos, California

Rocket Science

Purpose
Students learn about Newton's third law of motion and the basics of rocketry.

Time Required
20–25 minutes

Lab Ratings

EASY ————————————→ HARD

TEACHER PREP

CONCEPT LEVEL

CLEAN UP

MATERIALS

- index card (5 x 8 in.)
- film canister with a lid that seals on the in-side—do not use canisters with lids that close around the outside
- transparent tape
- sheet of construction paper
- scissors
- small jar of tap water
- seltzer tablet

Safety Information
The rocket may take 15–20 seconds to build up enough pressure to launch. Do not approach the rocket prematurely.

Wear safety goggles when launching the rocket.

What to Do

1. Tell students that in this demonstration you are going to prepare and launch a model rocket.

2. Roll the index card to form a tube 3 cm in diameter, and then slide the canister inside. Align the lid of the canister with one end of the index card, and securely tape the canister and the tube together. (The activity will not work as well if the paper and canisters come apart.)

3. Seal the tube by taping the seam where the edges of the card meet.

4. Cut two triangular paper fins from the construction paper. Attach them to the rocket just above its base and just below its midpoint. Explain to students that the fins act to stabilize the rocket in flight. Make a small cone of paper, and attach it to the top of the cylinder with tape to create the rocket's nose.

5. Take students outside to launch your rocket. If it is windy, place a quarter in the film canister to keep the rocket from blowing over.

6. Add enough water to fill approximately one-quarter of the canister.

7. Cut the seltzer tablet in half, and add one half to the water. QUICKLY snap the lid in place.

8. Set the rocket lid-side down on the ground. Step back, and prepare for liftoff! Rockets can shoot more than 5 m into the air.

continued...

Richard Petitt
Bardstown Middle School
Bardstown, Kentucky

EARTH SCIENCE

Explanation

You have built a rocket in its simplest form. The escaping gas at one end of the rocket provides the thrust needed to propel the rocket. This rocket works by the same principles as the massive vehicles that travel into space.

Newton's third law of motion states that whenever one object exerts a force on a second object, the second object exerts an equal and opposite force on the first. In the case of a rocket, the engine exerts a force to expel gases. In turn, the gases exert a force on the engine, pushing the rocket upward. A similar pair of forces is utilized in the model rocket. The canister exerts a downward force on the gas. (You don't notice this force on the gas because the lid prevents the gas from moving downward.) The gas exerts an upward force on the canister, pushing the rocket upward away from the lid.

Discussion

Use the following questions as a guide to encourage class discussion:

• What happened inside the canister? (As the seltzer tablet dissolved, carbon dioxide gas was released. This accumulating gas increased the pressure inside the canister. When that pressure became high enough, it popped the lid off the canister and the rocket moved upward.)

• What forces are exerted at the same time in this demonstration? (The mass of the rocket pushes on the escaping gas, and the escaping gas pushes on the rocket.)

• When does the rocket have the power to lift off? (The rocket lifts off when the upward push of the gas overcomes the inertia of the rocket.)

• Why did the rocket stop moving upward? (When there was no longer any upward push provided by the gas leaving the canister, there was no force to prevent gravity from pulling the rocket downward, so the rocket fell.)

MAKING MODELS

The Dollar Bill Bridge

Purpose

By supporting a drinking glass on top of a dollar bill, students are introduced to the importance of modeling in science.

Time Required

5–10 minutes

Lab Ratings

EASY ——————————→ HARD

TEACHER PREP

CONCEPT LEVEL

CLEAN UP

MATERIALS

- 3 lightweight drinking glasses (made of glass)
- metric ruler
- crisp, new dollar bill

What to Do

1. Place two drinking glasses 10 cm apart on a table.

2. Tell students that you will use the dollar bill to form a bridge that is strong enough to support the third glass. Ask students if they think this is possible. *(Expected answer: No.)*

3. Explain that it *is* possible, and encourage students to brainstorm for solutions.

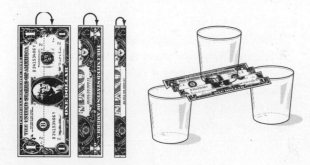

4. Fold the dollar bill in half *lengthwise*, then fold it in half lengthwise twice more. Open the bill, and fold it alternately along the creases, using your fingernail to press the creases. The folds should create a zigzag pattern, as shown below.

5. Spread the pleats out slightly, and place the dollar bill so that it bridges the two glasses. Carefully place the third glass on top of the bill and in the center.

Explanation

Paper is typically a poor material with which to build a bridge because it is so pliable. However, by changing the shape of the dollar bill, its rigidity and strength are increased tremendously. This example can be used to introduce the importance of using models in science. With a properly constructed model, engineers are able to test their design ideas before actually building a structure. This process saves time and money and is much safer.

Discussion

Use the following questions as a guide to encourage class discussion:

- What made the dollar bill strong enough to support the glass? *(The pleats increased the rigidity of the dollar bill and helped to distribute the weight of the glass over a greater area, allowing the dollar bill to hold more weight.)*

- **Critical Thinking** How does this design make corrugated mailing boxes stronger? *(The corrugations increase the rigidity of the mailing boxes, just as the pleats did for the dollar bill.)*

Robert Bartholemew
Berkshire Jr.–Sr. High
Canaan, New York

PHYSICAL SCIENCE

Curious Cubes

Purpose
Students attempt to solve the mystery of the sunken ice cube, and in the process they discover how relative densities determine buoyancy.

Time Required
10–15 minutes

Lab Ratings

EASY ———————————→ HARD

TEACHER PREP

CONCEPT LEVEL

CLEAN UP

MATERIALS

- 250 mL beakers (2)
- 200 mL of tap water
- food coloring
- 200 mL of rubbing alcohol or a nontoxic liquid with a density less than 0.92 g/cm³
- rubber gloves
- 2 ice cubes

What to Do

1. Fill one beaker with 200 mL of water. Fill the other with 200 mL of rubbing alcohol.

2. Add two drops of food coloring to each beaker to improve visibility.

3. Ask a student wearing safety goggles and rubber gloves to place an ice cube in each of the beakers.

4. Ask students to observe what happens. *(The cube in the alcohol will sink, while the cube in the water will float.)*

Explanation
The density of water is 1.00 g/cm³. The density of ice is 0.92 g/cm³. The density of rubbing alcohol is 0.79 g/cm³. Therefore, ice is more dense than alcohol and less dense than water. Ice floats on water and sinks in alcohol because a substance floats only if it is less dense than the fluid that surrounds it. Water is one of the few substances that is less dense in its solid form than as a liquid. As water freezes, its crystalline structure becomes more open and the ice expands, making it less dense.

Discussion
Review the concept of density with students. Explain that a cork floats because it is less dense than water, not because it is lighter. Encourage class discussion using the following questions as a guide:

- Why did one ice cube sink while the other floated? *(The ice cube was more dense than the rubbing alcohol, so it sank. It floated in the water because it was less dense than the water.)*

- How does the fact that water expands as it freezes explain the floating ice cube? *(Because the same mass of water occupies a greater volume when it freezes, the ice cube must be less dense. This is why it floats in liquid water.)*

Robert Bartholemew
Berkshire Jr.–Sr. High
Canaan, New York

DISCOVERY LAB

The Dancing Toothpicks

Purpose
Students learn about surface tension.

Time Required
5–10 minutes

Lab Ratings
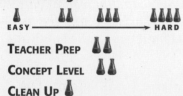
EASY ———————→ HARD

TEACHER PREP

CONCEPT LEVEL

CLEAN UP

MATERIALS

- dinner plate
- tap water
- 6–8 toothpicks
- sugar cube or any other absorbent material
- small piece of soap

What to Do

1. Fill the plate with water almost to the rim.

2. Arrange 6–8 toothpicks in a radial design in the middle of the plate. Do not allow any of the toothpicks to touch each other or the edge of the plate.

3. Ask students what they think will happen when you touch the water with a sugar cube. (*Accept all reasonable responses.*)

4. Touch the water in the center of the circle with the sugar cube. Notice that all of the toothpicks move toward the sugar cube.

5. Now ask the students what they expect to happen when you touch the water with a piece of soap. Then use a small piece of soap to touch the water. This time, all of the toothpicks will move away from the soap.

Explanation

When a sugar cube is placed among the toothpicks, the toothpicks move toward the cube. This is an illustration of capillary action. Because the water at the surface is absorbed by the sugar, a small current is created and the toothpicks are carried along the surface toward the sugar.

The molecules of water are attracted to one another due to intermolecular forces. Molecules on the surface of the water are only attracted by the molecules below and beside them. This is the reason droplets are spherical. The surface molecules are pulled in by the intermolecular forces such that they form the shape with the smallest surface area—a sphere.) The molecules of water are resistant to an arrangement that will increase the water's surface area. This resistance is called surface tension.

When the soap is dipped into the water, the surface tension is reduced at the point where the soap touched the water's surface. The surface tension of the water elsewhere is now relatively greater—an unbalanced force. The water molecules at the surface are pulled towards the region of higher surface tension, and the toothpicks follow.

continued...

Robert Bartholemew
Berkshire Jr.–Sr. High
Canaan, New York

PHYSICAL SCIENCE

Although the toothpicks move in opposite directions in these two demonstrations, the operating principle is the same. The toothpicks move because they are suspended on the water's surface, and moving the molecules at the water's surface moves the toothpicks. When water is pulled into the sugar cube, the molecules of water on the surface move toward the center. When soap is used to reduce the surface tension at the center, the molecules of water at the surface—and the toothpicks—are pulled away from the center.

Discussion

Use the following questions as a guideline to encourage class discussion:

- Why do the toothpicks move toward the sugar cube? *(By absorbing some of the water, the sugar cube creates a current that pulls the toothpick in.)*

- Why do the toothpicks move away from the soap? *(As the soap film reduces surface tension the water's surface— and the toothpicks suspended on it—are pulled outward.)*

DISCOVERY LAB

Does 2 + 2 = 4?

Purpose

Students learn how the particle theory of matter can be used to explain a surprising change in volume when two liquids are combined.

Time Required

5–10 minutes

Lab Ratings

EASY ———————————→ HARD

TEACHER PREP

CONCEPT LEVEL

CLEAN UP

MATERIALS

- 100 mL graduated cylinders (2)
- rubbing alcohol
- tap water
- marbles, enough to fill a $\frac{1}{2}$ cylinder
- sand, enough to fill a $\frac{1}{2}$ cylinder

What to Do

1. Pour 50 mL of rubbing alcohol into a graduated cylinder.

2. Pour 50 mL of water into a second graduated cylinder.

3. Have a student carefully measure the two volumes and write the results on the chalkboard.

4. Ask students to predict what the total volume will be when the two liquids are combined. *(Expected answer: 100 mL)*

5. Pour the contents of one cylinder into the other. Have another student measure the new volume. It should measure between 96 and 98 mL. Ask students to form a hypothesis about what happened to the missing liquid.

Explanation

Tell students that compounds are substances made up of more than one element that are chemically combined. Water and alcohol are compounds. The smallest particle of a compound that still retains the properties of that compound is called a molecule. Molecules are often irregularly shaped. Because of this, a beaker full of molecules is also a beaker full of tiny spaces between those molecules, just as in a full cereal bowl there are spaces between the cornflakes. Even liquids such as water and alcohol, which appear to have no spaces, are full of tiny molecular spaces. Some of the molecules of water fit into the spaces between the alcohol molecules.

When the liquids are combined, their total volume is less than the sum of their individual volumes because alcohol molecules and water molecules have a different shape and size. Although the two substances do not combine chemically, they do combine completely and uniformly. The spaces that exist between the molecules of each liquid are filled by the molecules of the other liquid.

To further demonstrate this principle, repeat the process using the sand and marbles. Tell the students that the sand represents molecules of water and the marbles represent molecules of alcohol. Just as the sand finds space between the marbles, the water molecules find spaces between the alcohol molecules.

This demonstration helps illustrate the particle model of matter, which states that all matter is composed of smaller particles.

continued...

Tracy Jahn
Berkshire Jr.–Sr. High
Canaan, New York

PHYSICAL SCIENCE

Discussion

As students attempt to solve the mystery, they may come up with the following hypotheses:

- The missing liquid remains in the original container. *(This explanation can be tested by using two flasks of water and repeating the process described above. This time there should be virtually no loss of liquid.)*

- The missing liquid evaporated during the experiment. *(Students may observe the volume of the combined liquids at 5-minute intervals to determine if evaporation can occur after the initial combination. Evaporation does not occur quickly enough to explain the loss of liquid.)*

Use the following questions as a guideline to encourage class discussion:

- Why didn't 50 mL of alcohol and 50 mL of water yield 100 mL of liquid when combined? *(The two substances have spaces in them, and those spaces provide room for some of the water molecules to settle among the alcohol molecules.)*

- Can you see the spaces in the liquids? *(No, the spaces exist at the molecular level, which is too small for us to see.)*

- How does this demonstration confirm the particle theory of matter? *(This demonstration provides evidence that there is considerable space in a beaker of liquid. This space could exist if a beaker of liquid consisted of tiny particles with spaces among them.)*

Demonstration with a CRUNCH!

Purpose
Students learn about changes in state as they watch popcorn pop.

Time Required
20–25 minutes

Lab Ratings
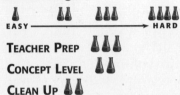

EASY ———————————→ HARD

TEACHER PREP 🧪🧪🧪
CONCEPT LEVEL 🧪🧪
CLEAN UP 🧪🧪

MATERIALS

- 500 mL of popcorn kernels*
- paper towels
- 2 containers
- masking tape
- felt-tip marker
- hot-air popcorn popper

*This lab works best with non-gourmet popcorn.

Advance Preparation
Divide the popcorn kernels into two 250 mL batches. Soak one batch in water for one hour. After soaking, blot the kernels with paper towels to remove excess water. Heat an oven to 200°F, and dry the second batch of kernels in it for one hour. Use a piece of tape and a marker to label one container "Dried popcorn" and the other "Soaked popcorn." Put the kernels in their respective containers.

You may wish to have salt and butter available to flavor the popcorn so you and your students can enjoy it after the experiment.

What to Do
1. Explain that you have two batches of popcorn kernels. One was soaked in water overnight, and the other was dried in an oven. Ask students to predict which batch will pop better.

2. Pop the oven-dried popcorn in the hot-air popper according to the manufacturer's directions. Ask students to describe what they observe. *(Few kernels popped.)*

3. Now pop the water-soaked popcorn kernels. First be sure they are completely dry. Ask students to describe what they observe. *(Many kernels popped.)*

Explanation
The inside of a kernel of popcorn is filled with starch. This starch expands into the white, fluffy material we call popcorn. The expansion occurs because a small amount of water in the kernel evaporates as the kernel is heated; the water expands as it changes into a gas, just as all gases expand when they are heated.

The batch that you soaked in water should have contained slightly more water than average fresh popcorn contains. Soaking the kernels ensured that you had good popping results. The batch that you dried should have contained less moisture; therefore, it was similar to old, stale popcorn. This process ensured that not as many popcorn kernels popped since there was not as much moisture available to expand into steam.

Discussion
Why did the soaked batch pop better than the dry batch? *(Popcorn pops because the water inside the kernel expands as it changes to steam. When the batch was dried, this water evaporated. When it was put in the popper, there was not enough water to create the pressure to make popcorn.)*

Tracy Jahn
Berkshire Jr.–Sr. High
Canaan, New York

PHYSICAL SCIENCE

Dense Suspense

Purpose

Students observe a can of regular soda sink in water and a can of diet soda float to learn about the concept of relative density.

Time Required

5–10 minutes

Lab Ratings

EASY ——————————————→ HARD

TEACHER PREP

CONCEPT LEVEL

CLEAN UP

MATERIALS

- large, deep, transparent bowl
- 2 unopened soda cans; one diet soda, one regular soda
- tap water
- dime (or some other small, dense object lighter than a soda can)

Advance Preparation

Before students enter the classroom, place the soda cans in the bowl. Add enough water to show that one floats and the other sinks.

What to Do

1. Show students that the can of diet soda is floating at the top of the bowl and that the can of regular soda has sunk to the bottom.

2. Ask students if they can explain why this happens. *(Accept all reasonable responses)*. They may suggest that the can of regular soda is heavier than the can of diet soda. This is true, but the cans' weight does not explain why they sink or float.

3. To illustrate this point, add a dime (or any other small, light object that will sink) to the water. The dime is lighter than the can of diet soda, but it sinks.

Explanation

An object will only float in a liquid if its density is lower than that of the liquid. Density is a measure of the amount of matter contained in a given volume. In other words, density is equal to an object's mass divided by its volume. The can of soda with sugar is denser than water, so it sinks. The can of sugar-free soda is less dense than water, so it floats.

Discussion

Use the following questions as a guide to encourage class discussion:

- Why did the dime sink in the water? *(The dime is denser than the water.)*

- **Critical Thinking** A submarine draws water into special tanks to sink and expels water in order to float. Using what you learned in this activity, explain how this process works. *(When the submarine draws the water into the tanks, it increases its mass without changing its volume. Therefore, its density increases and it sinks. When the submarine releases water, its density decreases and it floats.)*

Tracy Jahn
Berkshire Jr.–Sr. High
Canaan, New York

Newton's Eggciting Experiment

Purpose

Students are introduced to Newton's first law of motion.

Time Required

10–15 minutes

Lab Ratings

EASY ——————→ HARD

TEACHER PREP
CONCEPT LEVEL
CLEAN UP

MATERIALS

- 3 drinking glasses or clear plastic cups
- tap water
- pie pan
- 3 cardboard bathroom-tissue tubes
- 3 eggs
- broom

What to Do

1. Fill the glasses three-quarters full of water.

2. Arrange the glasses in a triangle near an edge of the table, as shown in the illustration at right. Set the pie pan on top of the glasses so that one edge of the pie pan extends over the edge of the table. If necessary, reposition the glasses. Place a cardboard tube vertically over each glass. Then place an egg sideways on top of each tube.

3. Stand the broom at the edge of the table so that the handle touches the edge of the pie pan. Carefully step on the bristles of the broom to hold it in place.

4. Tell students that you will knock the edge of the pie pan with the broom. Challenge them to predict what will happen. *(Accept all reasonable responses.)*

5. Pull the broom handle toward you, and then release it.

6. The broom handle should spring forward, driving the pie pan and the tubes out from under the eggs. The eggs should drop into the glasses of water directly below them.

Discussion

Have students work together in small groups to come up with an explanation of what happened. Suggest that they use Newton's first law, which states that an object at rest remains at rest and an object in motion remains in motion at constant speed and in a straight line unless acted upon by an unbalanced force.

Explanation

The broom applied force to the pie pan. When the edge of the pie pan struck the tubes, some of the force was transferred to the tubes. The tubes did not transfer enough of this force to the eggs to overcome their inertia. Therefore, the eggs remained in place until their support was removed. Gravity pulled the eggs downward until the water cushioned their fall and absorbed their downward energy.

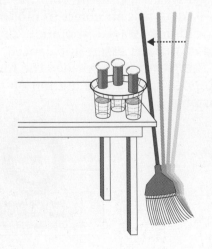

CLASSROOM
TESTED & APPROVED

Harriet Knops
Rolling Hills Middle School
El Dorado, California

PHYSICAL SCIENCE

Inertia Can Hurt Ya

Purpose

Students study a model collision to learn about Newton's first law of motion and the importance of seatbelts.

Time Required

15 minutes

Lab Ratings

EASY ———→ HARD

TEACHER PREP

CONCEPT LEVEL

CLEAN UP

MATERIALS

- 1.5 m board
- 5–10 books
- small toy cart with a removable passenger
- 20 cm piece of thread

What to Do

1. Place one end of the board on a pile of five books. Place two books at the other end of the board.

2. Place the passenger in the cart, and hold the cart at the top of the ramp. Ask students what will happen when you release the cart. *(Expected answer: The cart will roll to the bottom of the ramp and hit the books.)*

3. Release the cart. As the cart hits the books, the toy passenger will fly into the air. Ask students to explain why the figure flew through the air. *(An undisturbed object in motion tends to stay in motion, while an undisturbed object at rest tends to stay at rest.)*

4. Tell students that you are going to send the cart down the ramp again but that this time you want to be sure the passenger stays in the cart. Then use the thread to fashion a seat belt that will secure the passenger to the cart.

5. Place the cart at the top of the ramp, and release it again. This time, the passenger should stay in the cart.

Discussion

Use questions such as the following as a guide to encourage class discussion:

- Why did the toy passenger fly out of the cart the first time the cart was released down the ramp? *(The passenger in the cart was in motion, and an object in motion tends to stay in motion.)*

- Why did the passenger remain in the cart on the second run? *(Both the cart and the passenger are objects in motion. The books provide the unbalanced force that stops the motion of the cart, and the seat belt provides the unbalanced force that arrests the motion of the passenger.)*

Explanation

Newton's first law of motion states that an object at rest remains at rest and an object in motion remains in motion at constant speed and in a straight line unless acted upon by an unbalanced force. When the cart hit the books, the passenger in the cart was in motion. Although the cart stopped suddenly, the passenger stayed in motion due to inertia.

When a restraint was placed on the passenger and the cart crashed again, the passenger was not thrown from the cart. The passenger was an object in motion and would have stayed in motion, but instead it was met by an unbalanced force. The "seat belt" provided the force that stopped the forward motion of the passenger.

Brian Burnight
Big Bear Middle School
Big Bear Lake, California

DISCOVERY LAB

Fountain of Knowledge

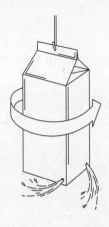

Purpose

Students observe the effects of Newton's third law as they set a simple water turbine in motion.

Time Required

15–20 minutes

Lab Ratings

EASY ——————————→ HARD

TEACHER PREP

CONCEPT LEVEL

CLEAN UP

MATERIALS

- empty milk carton
- 1–2 m length of sturdy string or fishing line
- nail
- pitcher filled with tap water
- bucket or sink

What to Do

1. Show students the milk carton, string, and nail. Challenge them to help you figure out how to make the carton spin on its own while it is suspended by the string. Then demonstrate one method for making the carton spin.

2. Use the nail to poke a small hole in the lower right-hand corner of each of the four vertical sides of the milk carton. Poke another hole through the top of the carton, as shown in the diagram.

3. Tie the string through the hole in the top of the milk carton, and hang the carton over the bucket or sink.

4. Fill the carton with water. As the water pours through the holes, the carton will begin to spin.

Explanation

Newton's first law states that an object at rest remains at rest unless acted on by an unbalanced force. Newton's third law states that all forces act in pairs. The carton exerts a force on the water, and the water exerts an equal and opposite force on the carton, except where the holes are placed. At the holes, gravity provides an unbalanced force on the water, causing it to move out through the holes. The movement of the water through the holes causes the carton to spin in the opposite direction of the water flow.

Discussion

Use the following questions as a guide to encourage further discussion:

- Newton's third law states that whenever one object exerts a force on a second object, the second object exerts an equal and opposite force on the first. What object is exerting a force on the water? *(the carton)* What object is exerting a force on the carton? *(the water)* What unbalanced force causes the water to move through the holes? *(gravity)*

CLASSROOM TESTED & APPROVED

Jane Lemons
Western Rockingham
Middle School
Madison, North Carolina

PHYSICAL SCIENCE

DISCOVERY LAB

The Rise and Fall of Raisins

Purpose

As students observe the effect of seltzer water on raisins, they learn about combined relative density.

Time Required

10–15 minutes

Lab Ratings

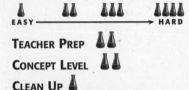

EASY ———————➤ HARD

TEACHER PREP

CONCEPT LEVEL

CLEAN UP

MATERIALS

- large, tall transparent jar or glass
- 5–10 raisins
- seltzer tablet
- tap water

What to Do

1. Fill the jar or glass with water. Drop a seltzer tablet into the water, and then drop the raisins in the water. Ask students to observe what happens to the raisins.

2. Continue observing the behavior of the raisins for a few minutes. Again, ask students to describe their observations. *(The raisins rise to the surface after a short time and then sink again. This process continues as long as the water fizzes.)*

3. Ask students to predict whether the raisins would exhibit the same behavior without the seltzer tablet. Empty the glass or jar and refill it with water. Drop the raisins into the water. Ask students: What happens? *(Without the seltzer tablet, the raisins sink to the bottom of the jar and stay there.)*

Explanation

The density of the raisins does not change. However, the carbon dioxide bubbles that adhere to a raisin's surface cause a much lower combined density of the raisin and bubbles. The bubbles add little mass to the raisin, but they displace an additional volume of water, causing the raisins with bubbles attached to them to float.

Discussion

Use the following questions as a guide to encourage class discussion:

- What was the purpose of the seltzer tablet? *(The tablet provided the bubbles that caused the raisins to rise and sink.)*

- Water is denser than carbon dioxide but less dense than a raisin. How can this statement be used to explain the behavior of the raisins? *(A raisin by itself is denser than water and therefore sinks, but a raisin covered with many carbon dioxide bubbles rises. This is because the combined density of the raisin and carbon dioxide bubbles is less than the density of water. At the surface, some of the bubbles burst. When this happens, the combined density of the raisin and carbon dioxide bubbles increases, and the raisin begins to sink again.)*

- **Critical Thinking** Use what you have observed in this activity to explain how a lava lamp works. *(As the wax reaches the warm light, its density decreases and the wax rises. As it floats away from the heat source, the wax cools and condenses. The density increases and the wax sinks again.)*

CLASSROOM TESTED & APPROVED

Brian Burnight
Big Bear Lake Middle School
Big Bear Lake, California

DISCOVERY LAB

Going Against the Flow

Purpose
Students learn how air pressure affects the flow of fluids.

Time Required
5–10 minutes

Lab Ratings

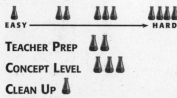

EASY ——————————————→ HARD

TEACHER PREP 🧪🧪

CONCEPT LEVEL 🧪🧪🧪

CLEAN UP 🧪

MATERIALS

- plastic bottle
- one-hole stopper that fits snugly into the plastic bottle
- scissors or a nail
- bucket or sink
- tap water

What to Do
1. Place the stopper firmly in the mouth of the plastic bottle. It is important that the fit be snug so that no air can enter or exit the bottle between the stopper and the mouth of the bottle. Use scis-

sors or a nail to puncture a small hole in the side of the bottle, close to the bottom. Ask students: What will happen when I fill the bottle with water? *(Expected answer: Water will come out of the hole at the bottom of the bottle.)*

2. Holding the bottle over a bucket or sink, fill the bottle with water. Keep the stopper unplugged, and while the water streams out of the bottle, ask students: How can I stop the stream of water without getting wet? *(Expected answer: It is impossible.)*

3. Tell students that it's possible to stop the stream without touching it. Then plug the hole in the stopper at the mouth of the bottle. The water will stop flowing from the bottle.

Discussion
Use questions such as the following to encourage class discussion:

- How was I able to control the flow of water by covering the hole in the stopper? *(Fluids flow from an area of high pressure to an area of low pressure. The water stopped flowing when air could no longer enter the bottle; the air pressure exerted on the hole at the bottom of the bottle was equal to the low pressure exerted by the air trapped inside the bottle plus the pressure due to the weight of the water. This equal pressure kept the water from coming out of the bottle.)*

- **Critical Thinking** If you want to pour juice out of a large can quickly, you should punch two holes in the top of the can. How does this help? *(One hole allows air to enter the can so that the juice can flow out of the other hole.)*

CLASSROOM TESTED & APPROVED

Donna Norwood
Monroe Middle School
Monroe, North Carolina

PHYSICAL SCIENCE

Pull-ease, Please!

Purpose

Students observe how a simple machine can make work easier.

Time Required

5–10 minutes

Lab Ratings

⚗   ⚗⚗   ⚗⚗⚗   ⚗⚗⚗⚗

EASY ——————————→ HARD

TEACHER PREP ⚗

CONCEPT LEVEL ⚗

CLEAN UP ⚗

MATERIALS

- 2 identical brooms
- rope or strong cord, about 3 m long
- metric ruler or measuring tape

What to Do

1. Ask two students to hold the brooms horizontally about 50 cm apart.

2. Tie the rope or cord to one broom, and wrap it around the handles as shown.

3. Invite a third student to hold the free end of the rope. Ask the class: Do you think the third student can pull the brooms together while the other two are holding them apart? Why or why not? *(Expected answer: No; two students should be able to exert more force than one.)*

4. Tell the student holding the rope to pull on it. The two brooms will move together even while the other two students try to keep them apart.

Explanation

The brooms and rope act together to form a pulley system. A pulley is a simple machine that can be used to increase a force. The force of the student pulling the rope was multiplied by the number of points where the rope pulled on the broom. Therefore, the third student exerted a force that was six times that of each of the other students.

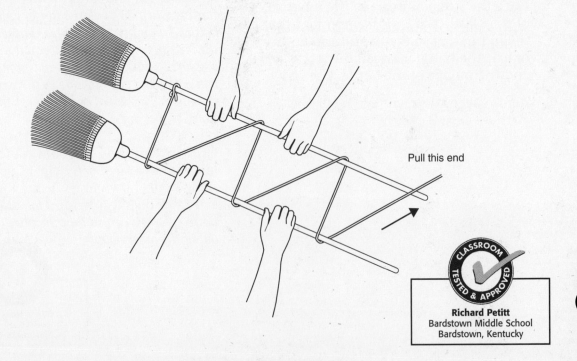

Pull this end

CLASSROOM TESTED & APPROVED

Richard Petitt
Bardstown Middle School
Bardstown, Kentucky

DISCOVERY LAB

A Clever Lever

Purpose

Students watch as a simple machine is used to lift a heavy piece of furniture.

Time Required

5–10 minutes

Lab Ratings

EASY ———————→ HARD

TEACHER PREP

CONCEPT LEVEL

CLEAN UP

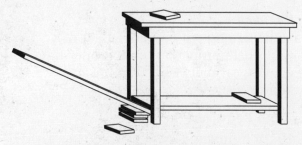

MATERIALS

- long wooden stick or metal pole, 3–4 m long
- 2–3 bricks or small wooden blocks
- heavy piece of furniture with a rung or horizontal bar near the bottom

What to Do

1. Ask students if they think you can lift a piece of heavy furniture with one finger. *(Expected answer: It is impossible.)*

2. Tell students that you know it is possible, and challenge them to come up with a method for doing so. Discuss the various hypotheses, and decide as a class which are the most feasible.

3. Test several hypotheses as a class. If students suggest using a lever, test that hypothesis last. If not, tell them that you will demonstrate a solution.

4. Place the blocks on the floor near the furniture rung. Insert the stick or pole under the rung and over the blocks, as shown in the diagram.

5. With one finger, press down on the other end of the stick. The furniture should lift easily.

Explanation

Explain to students that you used a simple machine called a lever to lift the object with one finger. In this experiment, the pole and bricks form a first-class lever. The fulcrum, or pivot, separates the effort and resistance. In this exercise, the furniture provided the resistance, and the pressure exerted by your finger was the effort. The blocks served as the fulcrum.

Students may be interested to know that pliers, scissors, crowbars, and seesaws are just a few examples of other levers.

Kristy Hollenbeck
Yutan Jr.–Sr. High
Yutan, Nebraska

PHYSICAL SCIENCE

Pendulum Peril

Purpose

Students observe the movement of a water-balloon pendulum to understand conversions between potential and kinetic energy.

Time Required

10–15 minutes

Lab Ratings

EASY ———————————→ HARD

TEACHER PREP

CONCEPT LEVEL

CLEAN UP

MATERIALS
• tap water • balloon • nylon string about 3 m long

What to Do

1. Attach the string to the ceiling. Fill the balloon with water and attach it securely to the free end of the string.

2. Adjust the length of the string so that the balloon just reaches your nose when you stand with your back against a wall. Securely tighten all of the knots.

3. Stand against the wall and hold the water balloon a few centimeters from your face. Ask students what they expect to happen when you release the balloon. *(Expected answer: It will swing forward and then swing back and hit you in the face. You will get soaked!)*

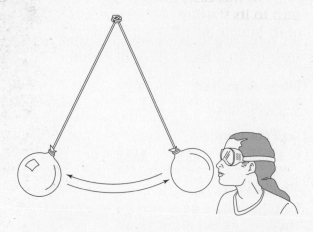

4. Now, with feigned anxiety, release the balloon and wait for it to swing back. Do not push the balloon or move your head!

Explanation

When the balloon is held in place, there is gravitational potential energy associated with its position. When the balloon is released, some of this potential energy is gradually converted into kinetic energy. At the midpoint of the swing, the kinetic energy is at a maximum, so the balloon is swinging at its fastest. After the midpoint of the swing, the kinetic energy is gradually converted back into potential energy. In the absence of friction, this cycle would continue indefinitely, and the balloon

continued...

Donna Norwood
Monroe Middle School
Monroe, North Carolina

would return to its starting position each time. In this case, the balloon does not return to its starting position because some of its kinetic energy is lost to friction. Explain to students that this energy is not destroyed. It is converted into another form of energy—heat. Point out that energy can be neither created nor destroyed.

The balloon does not swing beyond its original position because the total energy of the system cannot increase.

Discussion

Use the following questions as a guide to encourage class discussion:

- Why didn't the balloon hit me in the face? *(The balloon appeared to lose energy, which prevented it from returning to its starting position.)*

- What provided the energy for the balloon to move? *(The energy came from the balloon's position. This is called potential energy. When the balloon is released, this energy is converted, and the balloon moves.)*

- **Critical Thinking** Will the balloon have any energy when it stops swinging? Why or why not? *(Yes, the balloon still has potential energy because of its position. This energy would be released only if the string were cut.)*

Wrong-Way Roller?

Purpose

Students try to determine how an object appears to roll uphill, and in the process they learn something about the law of conservation of energy.

Time Required

5–10 minutes

Lab Ratings

EASY ——————————→ HARD

TEACHER PREP

CONCEPT LEVEL

CLEAN UP

MATERIALS
• masking tape • 2 identical plastic funnels • 2 metersticks • 2–4 books

Advance Preparation

Securely tape the large ends of the funnels together to make a roller. Place the two metersticks side by side on a table. Place one end of the sticks on a stack of books.

You may wish to practice this demonstration before class in order to determine how many books to use. The effect is more apparent when the metersticks form a gradual incline and when the separation between the metersticks is narrow.

What to Do

1. Ask students what happens when a round object is released on the side of a hill. *(It rolls downhill.)* Ask them why they think this is so. *(Accept all reasonable responses.)* Tell them that you will now cause a round object to roll uphill, defying the laws of physics.

2. Place the roller on the lower end of the two metersticks so that the joined ends of the funnels are between the two metersticks.

3. Grasp the raised ends of the two metersticks, one in each hand. Carefully slide the ends apart. The roller will roll toward you, seemingly uphill.

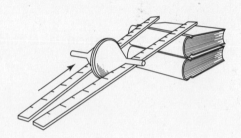

4. Ask students to discuss whether the roller is really violating any laws of physics. Then tell them that no laws of physics have been violated in this demonstration. Have them brainstorm a possible explanation for why the funnels appeared to roll uphill.

Explanation

The roller only appeared to move uphill. The roller is narrower at the ends than in the middle. As the gap between the metersticks widened, the roller actually moved downhill, as expected. This is easier to see by repeating the demonstration and asking students to carefully observe the path of the center of the roller in relation to the ground. They will discover that the center of the roller actually rolled downhill, as shown in the illustration above.

continued...

Kristy Hollenbeck
Yutan Jr.–Sr. High
Yutan, Nebraska

The law of conservation of energy states that in a closed system, the total energy does not change. The two types of energy in this demonstration are kinetic and potential. When the roller went from a resting state to a moving one, the roller translated some of its potential energy into kinetic energy. While the energy in the system remained constant, the roller gained kinetic energy and lost potential energy.

Discussion

Use the following questions as a guide to encourage class discussion:

• What happens to the potential energy of the roller when it moves down the sticks? *(Its potential energy decreases. Some of its potential energy becomes kinetic energy.)*

• Does the roller defy the laws of physics? *(No. It actually rolled downward, preserving the energy in the system. As the gap between the sticks became wider, the roller dropped farther down into it, though the angle of the sticks made it appear to roll upward.)*

• **Critical Thinking** How does this demonstration confirm the law of the conservation of energy? *(The roller could not move uphill and gain potential energy without energy being added to the system. It could only translate some of its potential energy into kinetic energy, which it did when it rolled downward.)*

PHYSICAL SCIENCE

DISCOVERY LAB

Cool It

Purpose
Students explore the concepts of temperature and evaporation.

Time Required
10–15 minutes

Lab Ratings

EASY ————————————→ HARD

TEACHER PREP
CONCEPT LEVEL
CLEAN UP

MATERIALS

- 2 identical indoor thermometers
- portable fan

What to Do

1. Place one thermometer 0.5 to 1 m in front of the fan, and place the other thermometer in the center of the room, out of the path of the fan. Do not turn on the fan.

2. Ask a student to record the temperature on each of the thermometers. The thermometers should give the same temperature reading.

3. Turn on the fan. Ask students to predict what will happen to the reading on the thermometer in front of the fan. *(Expected answer: The reading of the thermometer in front of the fan will be lower than the reading on the other thermometer.)*

4. After a few minutes, have the student read the thermometers again. Both thermometers should read roughly the same temperature as before you turned on the fan.

Discussion
Use the following questions as a guide to encourage class discussion:

- How does a fan create a breeze? *(As the blades turn, more air molecules are directed away from the fan, creating an air current.)*

- Why didn't the breeze lower the temperature of the air? *(The increased movement of air molecules does not lower the temperature of the air. In fact, an extremely sensitive thermometer might read a slightly higher temperature due to the increased movement of the air molecules.)*

- Why does the breeze feel cool? *(The breeze from the fan cools the skin by increasing the evaporation of sweat from the skin's surface. The evaporation cools the body.)*

CLASSROOM TESTED & APPROVED

Michael E. Kral
West Hardin Middle School
Cecillia, Kentucky

Candy Lights

Purpose

Students see that crushing some materials causes the materials to emit light, an effect known as *triboluminescence.*

Time Required

15 minutes

Lab Ratings

EASY ——————————→ HARD

TEACHER PREP
CONCEPT LEVEL
CLEAN UP

MATERIALS

- wooden or plastic cutting board
- roll of wintergreen candies
- hammer

Advance Preparation

This demonstration requires absolute darkness. If your classroom is difficult to darken completely, you may want to use an interior room, a large closet, or a dark storage area. You may want to try this demonstration in advance to be certain that conditions permit seeing the desired effect. Everyone should be wearing safety glasses during the demonstration.

What to Do

1. Place the cutting board on a stable surface, such as a lab bench, desk, or floor.

2. Place 3–4 candies on the cutting board.

3. Ask students to gather around the board so that they can all see the candies clearly.

4. Darken the room completely by turning off all lights and sealing the windows. Wait a few minutes for everyone's eyes to adjust to the darkness.

5. Ask students to watch the candies closely and observe what happens when the hammer hits them. Using the hammer, strike the candies several times.

6. Turn the lights back on. Encourage the students to discuss their observations and offer explanations.

7. Discuss the various hypotheses as a class.

Explanation

Students should see a flash of blue light as you crush the candy. This effect is due to charge separation between the pieces of candy and is called *triboluminescence.*

To help students understand what they have seen, you may wish to explain what happened as follows: When the wintergreen candy is crushed, the pieces of candy tend to be left with extra electrical charges on them. Some pieces are positively charged and therefore have fewer electrons than protons, and some are negatively charged and therefore have more electrons than protons. The extra electrons on the negatively charged pieces are free to move, so they flow to the positively charged pieces. The excess charges cancel each other out. This flow of charge is an electric current. As the electric current flows, it excites the atoms in its path, which then emit visible light in response. This is the light observed in the activity.

This activity can be compared to the generation of lightning in a thunderstorm. During a storm, excess charges tend to build up in different parts of a cloud or clouds. Eventually, the excess negative charge flows from one region to a region of excess positive charge. The electric current excites air molecules that emit the light we associate with a lightning bolt.

Jerry Taylor
Santa Maria Middle School
Phoenix, Arizona

PHYSICAL SCIENCE

As a Matter of Space

Purpose

Students model the helium atom and demonstrate that there is a vast amount of empty space in the atom. This activity can be used in conjunction with a discussion of the atomic models of Bohr and Rutherford.

Time Required

15–20 minutes

Lab Ratings

EASY ——————→ HARD

TEACHER PREP

CONCEPT LEVEL

CLEAN UP

MATERIALS

- metric measuring tape
- ball bearing
- 25 m lengths of string (2)
- 2 grains of sand

What to Do

1. Find an area large enough to create a circle with a diameter of about 50 m.

2. Tell students that they will model the structure of a helium atom. Choose a volunteer to hold the ball bearing and one end of both strings. Tell students that the ball bearing represents the nucleus of an atom.

3. Tell students that the grains of sand represent electrons. Choose two more volunteers to each hold an "electron."

4. Ask the two students carrying a sand grain to each pick up a different end of string and walk in opposite directions 25 m from the "nucleus."

5. Tell students that the distance between the ball bearing and each sand grain represents the relative distance between the nucleus of an atom and the first ring of electrons.

Explanation

In this activity, students have created their own atomic model based on the models of Ernest Rutherford and Niels Bohr.

In 1911, Ernest Rutherford proposed an atomic model. He bombarded thin gold foil with fast-moving, positively charged particles called alpha particles. To his amazement, most of the particles passed through the foil without being deflected. After studying the results further, Rutherford and his colleagues concluded that very few (one in 8,000) particles were deflected back toward the source. Rutherford concluded that an atom's structure is mostly empty space surrounding a dense, positively charged nucleus. He also concluded that the electrons orbit the nucleus in the same way that planets orbit the sun.

Rutherford continued studying the atom with Niels Bohr. In 1913, Bohr proposed his own model of the hydrogen atom. Bohr suggested that the electrons orbit in energy levels at discrete distances from the nucleus.

Going Further

Discuss the relative masses of protons and electrons using one grain of rice to represent an electron and 2,000 grains of rice (125 mL) to represent a proton.

Lee Yassinski
Sun Valley Middle School
Sun Valley, California

DISCOVERY LAB

Waiter, There's Carbon in My Sugar Bowl

Purpose

Students will draw a connection between organic matter and its fundamental element, carbon, as they view a spectacular chemical reaction. They will learn about chemical changes and prove that sugar contains carbon.

Time Required

10–15 minutes

Lab Ratings

EASY ⟶ HARD

TEACHER PREP 🔬🔬

CONCEPT LEVEL 🔬🔬

CLEAN UP 🔬🔬🔬🔬

MATERIALS

- 100 mL beaker
- 30 mL of sugar
- 5–10 mL of sulfuric acid
- a face shield
- chemical-resistant gloves
- tongs

Safety Information

Use extreme caution while performing this demonstration. Sulfuric acid is extremely caustic and can damage body tissue. The vapors can also cause severe eye and lung irritation and may cause tissue damage. The beaker and its contents get very hot, as well. Minimize risk by performing this demonstration under a fume hood or outside, and use protective goggles, a face shield, and gloves. Be sure students stay **at least** 1–2 m away from the beaker at all times. Handle the chunk of carbon produced in the experiment with tongs, as it will still be coated with sulfuric acid.

What to Do

1. Find a suitable location for this demonstration. Make sure there is enough room for the students to stand 1–2 m from the reaction and still see clearly.

2. Place 30 mL of sugar in the beaker.

3. Slowly add 5–10 mL of sulfuric acid to the sugar.

4. Stand back and watch! As clouds of steam and smoke are produced, students should observe carbon "growing" out of the beaker.

Explanation

When the sulfuric acid (H_2SO_4) combined with the sugar ($C_{12}H_{22}O_{11}$), a chemical change took place. As the sugar was dehydrated, thermal energy was released. Water vapor and sulfur dioxide escaped as gases. The black substance left in the beaker is carbon.

Discussion

Use the following questions as a guide to encourage class discussion:

- Does the substance remaining in the beaker resemble any substance you've seen before? *(The substance resembles coal.)* Explain that coal is a form of carbon.

- What ingredient used in the demonstration may have contained carbon? *(Sugar)* How do you know? *(All living things contain carbon. Sugar is the byproduct of a living thing, a plant.)*

- Did you observe a chemical or physical change? *(Chemical)* How do you know? *(A new substance was formed.)*

Brian Burnight
Big Bear Middle School
Big Bear Lake, California

PHYSICAL SCIENCE

DISCOVERY LAB

Fire and Ice

Purpose

By watching you use a chip of ice to start a fire, students learn about the differences between chemical and physical changes.

Time Required

20–25 minutes

Lab Ratings

EASY ———————————————→ HARD

TEACHER PREP
CONCEPT LEVEL
CLEAN UP

MATERIALS

- finely chopped tissue paper
- fire-resistant plate or tile
- 5 mL of sodium peroxide
- ice chips

Safety Information

Please note that this demonstration presents a serious risk of fire or explosion. Therefore, use extreme caution when performing this demonstration. To reduce the risks, use a small chip of ice, not an ice cube or any larger volume of water. You may wish to try this demonstration with an adult before students are present. The teacher and all students should stand clear of the reaction.

What to Do

1. Pile the tissue paper on the plate in the shape of a cone with a height of approximately 5 cm.

2. Ask students whether it is possible to burn the tissue paper with a chip of ice. (*Expected answer: It is impossible.*)

3. Place 5 mL of sodium peroxide on top of the cone of tissue paper as shown in the diagram.

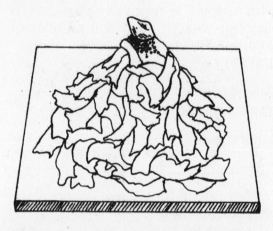

4. Hold up an ice chip for students to see. Then place the ice chip on top of the cone and step back. **Caution: Be certain to use a small chip of ice, not an ice cube or any larger volume of water.**

Explanation

As the ice melts, the water reacts with the sodium peroxide and releases oxygen and thermal energy ($Na_2O_2 + H_2O \rightarrow 2NaOH + O$ + thermal energy). The release of thermal energy causes the reaction to continue after the initial addition of water. The thermal energy then continues the decomposition of the sodium peroxide, releasing more oxygen. The oxygen released by this reaction is in *status nascendi*, which means that the oxygen is in its atomic form and has not combined with any other elements. Thus, the oxygen is very reactive. The oxygen reacts very quickly with the carbon in the paper. Tearing the tissue paper into small strips reduces the paper's kindling temperature and aids in starting the fire.

continued...

CLASSROOM TESTED & APPROVED

Alyson Mike
Radley Middle School
East Helena, Montana

Discussion

Use the following questions as a guide to encourage class discussion:

- What physical or chemical changes took place? How can you tell? *(In melting, the ice underwent a physical change; no new substances were formed. In reacting with the sodium peroxide, the water underwent a chemical change; new substances were formed. The sodium peroxide and the paper underwent chemical changes; a new substance was formed.)*

- What does the ice do in this reaction? *(It melts into water and initiates the release of the reactive oxygen from the sodium peroxide.)*

- What is required for burning to take place? *(Burning requires oxygen and a combustible substance. If the oxygen is in a stable form, the reaction also requires the energy necessary to combine with the combustible substance. The oxygen released in this demonstration is in a very unstable form, so no energy needs to be added to the reaction to start the fire; in fact, energy is released.)*

- **Critical Thinking** What could be used besides paper as a highly combustible substance? *(Examples include sugar, wood shavings, flour, straw, and coal dust.)*

- **Critical Thinking** What particular danger would be posed in fighting a fire in a laboratory that stored sodium peroxide? *(If water got into the sodium hydroxide, the two chemicals would react and this could cause the fire to spread. Fires involving sodium and other reactive metals are fought with class D fire extinguishers, which have no water in them.)*

Hoop It Up

Purpose

Students observe how the repulsion of like charges causes a Mylar™ hoop to hover over an aluminum pie pan.

Time Required

10–15 minutes

Lab Ratings

EASY ———————————→ HARD

TEACHER PREP
CONCEPT LEVEL
CLEAN UP

MATERIALS
• piece of Mylar™ foil, at least 0.7 x 35 cm • scissors • rubber cement • plastic-foam cup • aluminum pie pan • small piece of wool • plastic-foam plate

Advance Preparation

Mylar foil can be obtained from a first-aid kit or from most arts and crafts stores. Cut the foil to the proper size. When doing so, be sure the strip has a consistent width and a mass of 0.03 g or less. Any corners or jagged edges can disrupt the experiment. Join the two ends of the Mylar foil with rubber cement to form a closed loop. Glue the plastic-foam cup upside down on the inside of the aluminum pie pan, as shown in the sketch on this page.

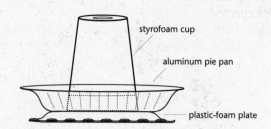

styrofoam cup

aluminum pie pan

plastic-foam plate

What to Do

1. Place the plastic-foam plate upside down on a table. Rub the plate with a piece of wool for about a minute. This will negatively charge the plate as electrons flow from the wool to the plate.

2. Place the aluminum pie pan upright on the plastic-foam plate. The negatively charged plastic-foam plate will attract positive charges to the bottom of the aluminum pie pan. The top of the aluminum pie pan will therefore have a negative charge.

3. Briefly touch the top of the pie pan with your finger. Negative charge will flow from the pie pan through your finger to the ground, leaving the pie pan with a positive charge.

4. Lift the pie pan by grabbing the plastic-foam cup. Turn the pan upside down and set the cup on the table. The pan is now ready for the demonstration.

continued...

Alyson Mike
Radley Middle School
East Helena, Montana

5. Hold the Mylar hoop above the pan, and ask students what they expect to happen when you drop the hoop. *(Expected answer: The hoop will fall and will either land on the pie pan or fall to the ground.)* Release the hoop.

6. The hoop will touch the pan and then rise to hover about 20–30 cm above the pan.

Explanation

When the hoop touches the aluminum pie pan, electrons from the hoop are attracted to the positively charged pan. Once electrons have flowed from the hoop to the pan, both the pan and the hoop are positively charged, so they repel one another. This electrostatic repulsion causes the hoop to rise and levitate above the pan.

Particles with opposite charges attract each other. Particles with the same charge repel each other. If these charges are strong enough, they can pull objects of opposite charges toward each other, like hair to a balloon. If two objects share the same charge, their repulsion can send them in different directions, like the hoop and the pan.

Discussion

Use the following questions as a guide to encourage class discussion:

• What did you learn about electrons from this demonstration? *(We know that electrons move. They physically move from the wool to the plate, from the pan to your finger, and through your body to the ground.)*

• When electrons leave the plate and are not replaced by other electrons, what happens to the charge on the plate? *(Electrons are negatively charged particles. When some of them leave the plate, the balance of charge shifts and the plate becomes positively charged.)*

• Why do the pan and the hoop repel each other? *(Both are positively charged, and like charges repel each other.)*

• Why does the hoop levitate? *(The force of gravity acts to hold the hoop on the pan. Because the repellent force between the positively charged hoop and pan is stronger than the force of gravity, the hoop levitates above the pan. These two forces are equal when the hoop is hovering.)*

DISCOVERY LAB

Bending Water

Purpose

Students learn how static electricity can bend a stream of water.

Time Required

5–10 minutes

Lab Ratings

🧪   🧪🧪   🧪🧪🧪   🧪🧪🧪🧪

EASY ———————————→ HARD

TEACHER PREP 🧪

CONCEPT LEVEL 🧪🧪🧪

CLEAN UP 🧪

MATERIALS

- water faucet
- plastic comb
- piece of wool cloth

What to Do

1. Turn on the water faucet so that the water runs in a cold and steady but slow and thin stream.

2. Ask students: How could I make the stream of water bend toward the comb without touching the water? *(Accept all reasonable responses, but do not provide students with the answer at this point.)*

3. Rub the comb vigorously with the wool cloth, and hold the comb near the stream of water. The stream of water should bend toward the comb.

4. Ask students: What made this happen? *(Expected answer: Rubbing the comb with the wool caused the water to be attracted to the comb. Some students may even respond that it was caused by static electricity. If so, ask them to explain how static electricity caused this to happen.)*

Explanation

When the comb was rubbed with the wool, electrons were rubbed onto the comb. This caused the comb to become negatively charged. This buildup of charges is known as *static electricity*. The water molecules have both positive and negative charges. The positive ends of the water molecules were attracted to the negatively charged electrons, causing the stream of water to move toward the comb.

Discussion

Use the following question as a guide to encourage class discussion:

Do you know of any other examples of static electricity? *(Possible responses: clothes clinging together after coming out of the dryer, getting a shock when touching a metal doorknob, hair sticking up after putting on a sweater.)*

CLASSROOM TESTED & APPROVED

Kenneth Horn
Fallstone Middle School
Fallstone, Maryland

DISCOVERY LAB

Magnetic Nails

Purpose

Students help you build a simple electromagnet and then brainstorm about how to increase the electromagnet's strength.

Time Required

15–20 minutes

Lab Ratings

EASY ———————————————→ HARD

TEACHER PREP

CONCEPT LEVEL

CLEAN UP

MATERIALS

- 2 D cells
- electrical tape
- 1 m of 20–22 gauge insulated wire
- wire cutters
- 4–5 paper clips
- 8-penny iron nail
- 6 V battery
- 16-penny iron nail

Advance Preparation

Tape the two D cells together in a continuous circuit so that the positive pole of the first cell is connected to the negative pole of the second cell. Remove 2 cm of insulation from both ends of the wire. Place the paper clips in a pile on the desktop. Arrange all of the materials listed for this lab so that they can be seen clearly by the class.

What to Do

1. Ask students to think about how they could make a magnet using only the materials on display. Test some of the hypotheses.

2. Demonstrate one solution by tightly coiling the wire 30 times around the 8-penny nail. Be sure to leave at least 10 cm of the wire free at both ends.

3. Using the tape, attach one end of the wire to one pole of the D cells.

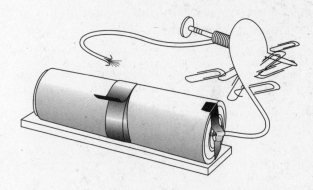

4. Hold the other wire against the other pole of the cells long enough to pick up some paper clips with the tip of the nail.

5. Lift some paper clips with the magnet. Hold them over a portion of the desktop, away from the pile of paper clips. Detach one wire from the cells, and let the paper clips fall.

6. Count the number of paper clips that fell, and ask the class to brainstorm about how to make the magnet pick up twice as many. Write suggestions on the board, and test the best ideas. *(One solution is to coil the wire more times around the nail. Alternately, you may use the 6 V battery or a bigger nail.)*

continued...

Kenneth Horn
Fallstone Middle School
Fallstone, Maryland

PHYSICAL SCIENCE

Explanation

This activity demonstrates that electrical current traveling through a wire creates a magnetic field. This field can be imagined as concentric circles around the wire. Coiling the wire increases the strength of the magnetic field tremendously because it closely aligns the fields in a parallel pattern. Using an iron core compounds the effect because the molecules of the core are temporarily lined up along the magnetic field. Increasing the voltage increases the current, which in turn creates a stronger magnetic field.

Discussion

Encourage further discussion with questions such as the following:

- What are some examples of electromagnets in your everyday life? *(Electric motors, stereo speakers, and power car-door locks are three examples.)*

- Why does coiling the wire increase the magnetic force? *(Coiling brings the magnetic fields into parallel alignment with each other.)*

Going Further

Lightly cover a glass plate with iron filings and place a simple bar magnet (not an electromagnet) under the plate. The filings will arrange into a pattern that helps illustrate the lines of magnetic force.

DISCOVERY LAB

Pitch Forks

Purpose

Students observe the nature of waves.

Time Required

10–15 minutes

Lab Ratings

EASY ———————→ HARD

TEACHER PREP

CONCEPT LEVEL

CLEAN UP

MATERIALS
• 3 tuning forks that produce different tones
• beaker
• tap water

Advance Preparation

To be sure that every student is able to see the activity, it may be necessary to set several beakers around the classroom. This will allow smaller groups to watch as you perform this demonstration.

What to Do

1. Strike the three tuning forks. As you are doing this, ask students to listen for three distinct tones.

2. Ask students if they can imagine what different sound waves look like as they travel through air.

3. Fill the beaker with cold water.

4. Strike each of the three tuning forks. Ask students how many distinct tones they hear. *(Each fork produces a different tone, so students should hear three distinct tones.)*

5. Strike one of the tuning forks, and touch the prongs lightly to the surface of the water. Ask students to observe the waves created on the surface of the water. Repeat this with the other two forks.

Explanation

Each fork produces a different tone and creates waves of different wavelengths in the water. Sound travels in air in the same way that ripples move through water. As the prongs of the tuning fork vibrate, they push on the air molecules around them. This energy is passed from one molecule to the next. As sound waves move through the air, air molecules are alternately compressed and spread apart, much like a wave moving in water. Both the water and sound move in *longitudinal waves*.

As students listened to the three tuning forks, they should have noticed that each fork emitted a tone at a different pitch. The pitch of a sound is related to its frequency, or the number of vibrations in a given amount of time. A sound with a higher pitch is a wave with a higher frequency. Therefore, the high-pitched tuning fork created waves in the water with the highest frequency. The low-pitched tuning fork produced fewer waves in the same amount of time.

Kenneth Horn
Fallstone Middle School
Fallstone, Maryland

PHYSICAL SCIENCE

DISCOVERY LAB

A Hot Tone

Purpose

Students learn how air moving through a pipe can produce sound.

Time Required

15–20 minutes

Lab Ratings

EASY ———————→ HARD

TEACHER PREP

CONCEPT LEVEL

CLEAN UP

MATERIALS

- several pieces of PVC pipe 5–6 cm in diameter and 60–150 cm in length
- pipe 12–13 cm in diameter and 1 m in length
- circular piece of mesh screen or hardware cloth 12–13 cm in diameter, cut to fit snugly inside the largest pipe
- Bunsen burner or propane torch
- ignition device, such as flint
- heat-resistant gloves

HELPFUL HINTS

Plumbing supply stores sometimes have scraps of PVC pipe available at reduced prices.

Safety Information

Use extreme caution with a lit torch. Wear protective eyewear, and do not point the flame toward other people or materials. Do not place the PVC pipe too close to the flame—PVC can melt and give off dangerous fumes. Students should stand clear of the demonstration area. Have an adult present to assist with this demonstration.

What to Do

1. Show students the pieces of pipe. Ask them how the pipes might be used to create sounds. *(Sample answer: The pipes could be hit or blown into to create sounds.)* Tell students that you are going to use the Bunsen burner or torch to produce sounds in the pipes.

2. To keep the air flowing uniformly in the largest pipe, insert the wire-mesh screen a few inches from the top so that it is lodged horizontally inside the pipe.

3. Use the ignition device to light the propane torch or Bunsen burner, and adjust it to the hottest setting.

4. If you are using the torch, have another adult hold the torch so that the flame is pointed toward the ceiling.

5. While wearing heat-resistant gloves, hold the shortest pipe vertically about 20–30 cm over the flame. Slowly lower the pipe over the flame until the pipe begins to emit a tone. Experiment to find the height that results in the loudest tone. Ask students to match the pitch by humming.

continued...

Alyson Mike
Radley Middle School
East Helena, Montana

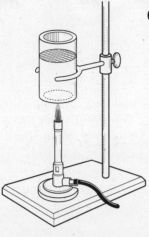

6. Repeat step 5 with successively longer pieces of pipe. Once you have finished with the longest piece of pipe, try step 5 with the 1 m pipe and the wire mesh. *(It should make a startlingly loud noise!)*

Pitch	Approximate length (cm)
C	132
D	118
E	104
F	98
G	88
A	78
B	70

Explanation

The flame heats the air directly above it, which expands and rises into the pipe. The hot air flowing through the pipe sets up a standing wave in the pipe. It is the motion of the air, caused by this standing wave, that produces the tone.

The pitch of the tone is determined by the length of the pipe. Although a pipe can produce a number of tones, in general it is easiest to produce the lowest possible tone of the pipe. The table at left gives the typical lengths of pipe necessary to produce the pitch in the left column, starting at a pitch one octave below middle C. These lengths are only approximate. Note that the shorter the pipe is, the higher the pitch it produces will be.

Discussion

Use the following questions as a guide to encourage class discussion:

- How is the tone produced? *(The hot moving gas in the tube sets up a standing wave, which produces sound.)*

- How does the pitch vary with the length of pipe? *(The longer the pipe is, the lower the pitch is. Shorter pipes create higher pitches.)*

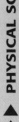

PHYSICAL SCIENCE

Jingle Bells, Silent Bells

Purpose
Students learn that sound waves must travel through a medium in order to be audible.

Time Required
20–25 minutes

Lab Ratings

EASY ———————————→ HARD

TEACHER PREP

CONCEPT LEVEL

CLEAN UP

MATERIALS

- 10 cm piece of wire clothes hanger
- small bell
- one-hole rubber stopper
- small amount of modeling clay
- boiling flask
- tap water
- hot plate
- watch or clock that indicates seconds
- heat-resistant gloves or oven mitts

What to Do

1. Bend the piece of hanger into a tight U-shape. Slide a small bell onto the wire so that the bell hangs at the curved portion of the U.

2. Slide the ends of the wire through the one-hole rubber stopper. Seal the hole around the wire with a plug of modeling clay.

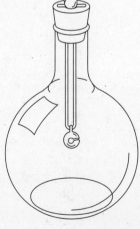

3. Boil a small amount of water in the flask, and then remove the

flask from the heat. Do not boil away the water.

4. Quickly insert the bell assembly into the flask. Don't let the bell touch the water. Let the flask cool for 1 minute. While wearing heat-resistant gloves, shake the flask. Tell students to listen carefully. Ask: Can you hear the bell? *(Students should hear only a faint sound, if any sound at all.)*

5. Loosen the stopper so that air can enter the flask. Then replace the stopper, and shake again. Ask: Can you hear the bell now? *(Students should be able to hear the bell clearly.)*

Explanation
The boiling water in the open flask produced steam, which expanded in the flask and displaced many of the air molecules. When the flask was plugged and removed from the heat source, the steam in the flask cooled and condensed back to a liquid. This left very few air molecules in the flask. When the bell was rung, the sound waves could not travel as easily with fewer air molecules to excite. As the clay plug was removed and replaced, air molecules returned to the flask. The sound vibrations were then able to travel, and therefore ringing could be heard.

Discussion
- What happened to the air in the flask when the water boiled? *(The steam forced the air out of the flask.)*

- What surrounded the bell in the flask after the air left? *(The bell was surrounded by a vacuum.)*

continued...

Kathy LaRoe
Radley Elementary School
East Helena, Montana

- Why didn't the bell make noise in a vacuum? *(Sound waves need a medium in order to travel. Without a medium, the vibration of the bell has no way of traveling.)*

- Why did the bell make noise after we opened the flask? *(Air entered the flask when we opened it, giving the sound waves a medium to travel through.)*

- **Critical Thinking** Dolphins communicate and locate objects in the ocean using sound waves. How can they make sound without air molecules? *(Sound can travel through any medium—solid, liquid, or gas. Sound waves only require molecules through which to move. When a dolphin sends sound waves through the water, the water molecules become excited in the same way that air molecules become excited as your voice travels through the air around you as you speak.)*

- **Critical Thinking** If astronauts were able to walk on the moon without spacesuits, would they be able to talk to each other? Hint: There is no air on the moon. *(Sound requires a medium to travel through. Astronauts would have to communicate with hand signals.)*

- Waves that require a medium through which to travel are called mechanical

waves. How did this experiment show that sound waves are mechanical waves? *(The bell could be heard more clearly when there was more air in the flask. This shows that sound waves require a medium through which to travel, so they are mechanical waves.)*

Going Further

Another way to illustrate that sound requires a medium in which to travel is to have students construct a telephone with two cans and a string. Each telephone will require two cans and a long piece of string. Have students punch a hole in the bottom of two cans. Have them tie a knot in one end of the string and thread it out the bottom hole of one can and into the bottom hole of another. Have them tie a knot in the other end of the string so that the two cans are joined securely. Ask a student to speak into one can while another student places his or her ear into the opening of the other. Experiment with different string lengths and tensions, different kinds of string, and no string at all.

PHYSICAL SCIENCE

MAKING MODELS

Hear Ye, Hear Ye

Purpose
Students learn how the eardrum works.

Time Required
10–20 minutes

Lab Ratings

A AA AAA AAAA
EASY ——————————————— HARD

TEACHER PREP AA
CONCEPT LEVEL AA
CLEAN UP A

MATERIALS

- soup can or coffee can
- can opener
- large balloon
- large rubber band
- bottle of glue
- mirror, about 2 × 2 cm
- flashlight, or laser pointer
- blank wall or projection screen

Advance Preparation
Remove both ends of the can with a can opener. Beware of the sharp edges. Cut a piece from the balloon that is large enough to stretch over one end of the can. Stretch the piece of balloon over one end of the can, and secure it tightly with a rubber band. Glue the mirror, slightly off-center, onto the outside of the balloon, and allow it to dry.

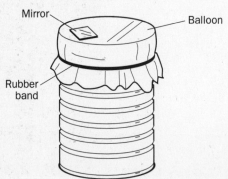

Mirror
Balloon
Rubber band

HELPFUL HINTS

You may wish to make two model eardrums so that students can compare the motion of the two images for different sounds.

What to Do

1. Ask students if they know how their eardrum works. *(Accept all reasonable answers.)* Tell them that you will use a model of an eardrum to demonstrate its function.

2. Shine the flashlight or laser pointer on the mirror so that the light reflects onto a wall or movie screen and can be seen by the entire class. You may wish to dim the lights in order to see the image more clearly.

3. Ask several students to speak loudly into the open end of the can one at a time. Alternatively, you may wish to use a tuning fork or musical instrument. Students should observe that the image vibrates as sound is produced in the can.

4. Encourage students to have fun making different sounds by playing different instruments or using different voices at different volumes and pitches. Challenge students to determine how the movement of the image changes in response to changes in the sound.

continued...

Kathy LaRoe
Radley Elementary School
East Helena, Montana

Explanation

The sound waves in the can cause the balloon to vibrate in much the same way that sound waves cause the eardrum to vibrate. The balloon will vibrate differently depending on the pitch and volume of the sound. This, in turn, will affect the reflection on the wall. The louder the sound is, the more the image will vibrate. The higher the pitch (frequency) of the sound is, the faster the image will vibrate.

Discussion

Use the following questions as a guide to encourage class discussion:

• Describe what happens to the image as different students speak into the model eardrum. *(The image vibrates. When one person speaks loudly into the model eardrum, the image vibrates a lot. When another person talks with a high-pitched voice, the light moves back and forth quickly.)*

• What causes the image on the wall to move? *(When a person talks into the model eardrum, sound waves strike the balloon. This makes it vibrate. When the balloon vibrates, so does the mirror. The image moves because the mirror is moving.)*

• What would happen to the image on the wall in each of these cases?

   a. You talk into the can very quietly. *(The image would not move much.)*

   b. Tweety Bird talks into the can. *(The image would vibrate very fast because Tweety Bird has a high-pitched voice.)*

   c. Darth Vader talks into the can. *(The image would vibrate very slowly because Darth Vader has a low-pitched voice.)*

PHYSICAL SCIENCE

The Sounds of Time

Purpose
Students observe the Doppler effect in this dramatic demonstration.

Time Required
5–10 minutes

Lab Ratings

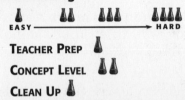

EASY — HARD

TEACHER PREP

CONCEPT LEVEL

CLEAN UP

MATERIALS

- alarm clock with handle
- piece of string, 1 m long

What to Do

1. Secure the end of the string to the top handle of the alarm clock. Make sure the string is tightly secured to the clock.

2. Make sure that you have enough room to swing the clock around you without hitting anyone.

3. Tell students that you are about to set off the clock's alarm and that when you do

so, you would like for them to listen carefully for changes in the sound.

4. Set off the alarm. Holding the other end of the rope, swing the alarm clock above your head.

Discussion
Use the following questions as a guide to encourage class discussion:

- Ask students: How does the sound change as the clock moves in a circle? *(Expected answer: The pitch of the sound increases and decreases as the clock moves in a circle.)*

- Why did you hear different pitches from the alarm? *(Sample answer: Although the alarm clock emitted a constant wavelength of sound, the waves were crowded as they reached me. Therefore, the observed frequency was higher in front of the clock and lower behind it.)*

continued...

Alyson Mike
Radley Middle School
East Helena, Montana

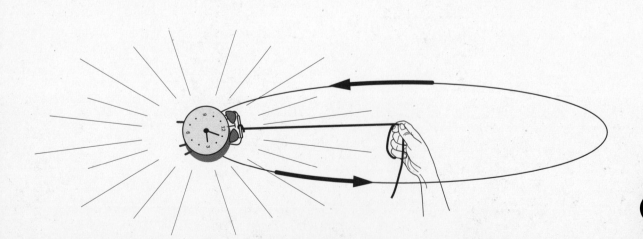

- Was the clock moving away or coming toward you when you heard the higher pitch? *(Sample answer: It was coming toward me.)*

- **Critical Thinking** If I were to use a shorter string and swing the alarm clock at the same rate over my head, what would you hear? *(Expected answer: The shift in pitch would decrease because the speed of the alarm clock would decrease.)*

Explanation

As the alarm clock swings in its circular path, it alternately moves closer to and farther from the listener's ear. As the alarm clock moves closer, more sound waves reach the ear in the same time interval. We call this a higher frequency or higher pitch. Conversely, when the clock is at its farthest point from the ear, fewer waves reach the ear, and we perceive this as a lower frequency or lower pitch.

Background Information

This demonstration uses sound waves to illustrate an important principle about waves. In 1842, Christian Johann Doppler made an interesting discovery. He observed that as an object producing a sound moved closer to him, the pitch of the sound increased. As the object passed him, the pitch decreased. This effect can be heard when a fire engine or a train approaches and passes by. Doppler suggested that the waves in front of a moving object are compressed and shortened, while the waves behind the object are longer. This phenomenon is called the *Doppler effect*.

The Doppler effect can be perceived in any moving object emitting energy in waves. For example, Doppler radar uses radio waves to track the speed of storms. Astronomers use the Doppler effect to study the motion of celestial bodies.

PHYSICAL SCIENCE

DISCOVERY LAB

Light Humor

Purpose

Students explore how their eyes adjust to different kinds of images.

Time Required

15–20 minutes

Lab Ratings

EASY ———————— HARD

TEACHER PREP
CONCEPT LEVEL
CLEAN UP

MATERIALS

- computer monitor
- neon night light
- incandescent light
- fluorescent light
- alarm clock with LED display (red numbers)

What to Do

1. Set up a viewing station for each light source. Turn off the power to all of the light sources.

2. Divide the class into groups at each viewing station.

3. Ask students to stick out their tongues and give a loud "raspberry" (also known as a Bronx cheer). In other words, ask them to make a rude buzzing sound with their lips.

4. Turn on the power to the light sources. Instruct students to stand facing their designated light source at a distance of 3–6 m. Have them look at the light source and make the loud noise again. The louder the better! Direct the students to take turns. One person at each station should experiment at a time.

5. Ask students to describe their observations in their ScienceLogs. What did they see? Was it what they expected?

6. Have the groups rotate to all of the stations, giving the "raspberry" at each.

7. After students have completed the activities at all of the stations, ask them questions about their observations. *(They should have observed that giving "raspberries" to the LED clock, the computer monitor, the neon light, and the fluorescent light caused the light sources to flicker. The incandescent light did not flicker.)*

8. Ask the students to hypothesize about the cause of the flickering. To give them a hint, ask them to place one hand on their heads and give another "raspberry." What did they feel? *(They should have felt their head vibrate.)*

Explanation

When a person views a neon light, a computer screen, an LED clock, or a fluorescent light under normal conditions, he or she does not perceive it to be flickering, but it is. These light sources flash on and off 60 or 120 times per second. Because the flashes are so short and fast, the human brain perceives the light to be constant.

However, when a person looks at the same light source and gives a "raspberry," the light appears to flicker. The person's eyes vibrate while he or she gives a "raspberry." As the eye moves, the image of the object being viewed is projected onto different portions of the retina. Under these conditions, the brain perceives the light to be flickering.

CLASSROOM TESTED & APPROVED

Donna Norwood
Monroe Middle School
Monroe, North Carolina